Lecture Notes in Computer Science 1818

Edited by G. Goos, J. Hartmanis and J. van Leeuwen

Springer
Berlin
Heidelberg
New York
Barcelona
Hong Kong
London
Milan
Paris
Singapore
Tokyo

Cambyse Guy Omidyar (Ed.)

Mobile and Wireless Communications Networks

IFIP-TC6/European Commission NETWORKING 2000
International Workshop, MWCN 2000
Paris, France, May 16-17, 2000
Proceedings

 Springer

Series Editors

Gerhard Goos, Karlsruhe University, Germany
Juris Hartmanis, Cornell University, NY, USA
Jan van Leeuwen, Utrecht University, The Netherlands

Volume Editor

Cambyse Guy Omidyar
Computer Sciences Corporation
7459-A Candlewood Drive
Hanover, Maryland 21076, USA
E-mail: gomidyar@csc.com

Cataloging-in-Publication Data applied for

Die Deutsche Bibliothek - CIP-Einheitsaufnahme

Mobile and wireless communication networks : European union
networking 2000 ; international workshop ; proceedings / IFIP TC 6,
MWCN 2000, Paris, France, May 16 - 17, 2000. Cambyse Guy Omidyar
(ed.). - Berlin ; Heidelberg ; New York ; Barcelona ; Hong Kong ;
London ; Milan ; Paris ; Singapore ; Tokyo : Springer, 2000
   (Lecture notes in computer science ; Vol. 1818)
   ISBN 3-540-67543-4

CR Subject Classification (1998): C.2, C.4, D.2, H.4.3, J.2, J.1

ISSN 0302-9743
ISBN 3-540-67543-4 Springer-Verlag Berlin Heidelberg New York

Springer-Verlag is a company in the BertelsmannSpringer publishing group.
© Springer-Verlag Berlin Heidelberg 2000
Printed in Germany

Typesetting: Camera-ready by author, data conversion by Boller Mediendesign
Printed on acid-free paper   SPIN 10720335   06/3142   5 4 3 2 1 0

Preface

Mobile and Wireless Communications Networks 2000 (MWCN 2000) was the second conference in its series aimed at stimulating technical exchange in the emerging and important field of Mobile and Wireless Communications Networks. The first conference took place at Sorbonne University, Paris, France May 12–15, 1997.

Integration of fixed and portable wireless access technologies and their mobility features into current fixed and mobile IP and ATM networks presents a cost effective and efficient way to provide seamless end-to-end connectivity and ubiquitous access in a market where demands on wireless networks have grown rapidly and which is predicted to generate billions of dollars in revenue. The deployment of broadband cell-based technologies including demand for internet mobility and their integration with emerging broadband wireless access networks (BWANs) are becoming increasingly important. Mobility and connection management, location management, connection establishment, quality of service provisioning, inter-working of wire-lines with wireless networks for voice, video, image, and data were some of the areas of focus in wireless networking of this conference. In the last few years with the development of digital technologies, satellites have emerged in the forefront of multimedia delivery techniques. From using current digital broadcast systems for web access to developing sophisticated platforms for integrated services, satellite networking is now a well-established solution for the broadband needs of the new millennium. This workshop was intended to provide a timely forum for exploratory research and practical contributions from North America, Europe, The Middle East, and The Far East. It helped designers to identify the theoretical and actual problems that face today's complex problems associated with mobile and wireless communications links and networks. This conference was sponsored by IFIP TC6, European Union, and Reseau National de la Recherche en Telecommunications (RNRT), in co-operation with ACM SIGMOBILE, the IEEE ComSoc Technical Committee on Communications System Integration and Modeling and took place at the Cité des Sciences, La Villette, Paris, France May 16–17, 2000.

March 2000   Guy Omidyar

Scientific Program Committee

Table of Contents

Mobility Management and Access Techniques
Chair: Mahmoud Naghshineh, IBM, Thomas J. Watson Research Center, USA

Mobility Support in IP
Chair: Kaveh Pahlavan, Worcester Polytechnic Institute, USA

An Overview of Wireless Indoor Geolocation
Techniques and Systems

Kaveh Pahlavan, Xinrong Li, Mika Ylianttila, Ranvir Chana, and Matti Latva-aho

Center for Wireless Information Network Studies, Worcester Polytechnic Institute, USA
{kaveh, xinrong}@ece.wpi.edu
Centre for Wireless Communications, University of Oulu, Finland
{over, rschana, matla}@ees2.oulu.fi

Abstract. Wireless indoor networks are finding their way into the home and office environments. Also, exploiting location information becomes very popular for both wireless service providers and consumers applications. However, the indoor radio channel causes challenges in extracting accurate location information in indoor environment so that traditional GPS and cellular location systems cannot work properly in indoor areas. This paper provides an overview of the indoor geolocation techniques. After introducing an overall architecture for indoor geolocation systems, technical overview of two indoor geolocation systems are presented. To demonstrate the predicted performance of such systems some simulation results obtained from an indoor geolocation demonstrator are presented.

1. Introduction

Today numerous wireless indoor products such as cordless telephone, wireless security systems, cordless speakers and even wireless Internet access have been introduced to the consumers. The same way as in late 70's and early 80's the increasing number of terminals in the offices initiated the LAN industry, today increasing number of wireless terminals in indoor areas is promoting wireless indoor networking. Indoor areas are difficult for wiring and most people are reluctant to allow workers to do extensive wiring inside buildings. Wireless is a cost efficient solution that can also provide additional feature of mobility and convenient relocatability. These reasons leave wireless as the preferred medium for indoor networking in the future and promote local networking activities such as Bluetooth, Home-RF, IEEE 802.11 and HIPERLAN/2.

An important evolving technology in recent years has been the indoor geolocation technology both for military and commercial applications. In the commercial application there is an increasing need for these systems for application in hospitals to locate patients or expensive equipment and in homes to locate children and equipment. Military and public safety applications in urban scenarios have promoted a need for inbuilding communication and geolocation networks enabling soldiers, policeman, and fire fighters to complete their missions in urban areas. These incentives have lead to research in indoor geolocation systems [1][2][3]. Due to indoor path loss, traditional GPS or E-911 location system cannot work properly in

C.G. Omidyar (Ed.): MWCN 2000, LNCS 1818, pp. 1-13, 2000.

indoor environment. As a result, dedicated indoor geolocation systems have to be developed to provide accurate indoor geolocation services.

This paper provides an update on the trends in indoor geolocation systems. In section 2, we briefly discuss the overall system architecture and various geolocation metrics that can be used in indoor environment. Then technical overviews of two indoor geolocation products are presented in Section 3. In Section 4, some simulation results obtained from an indoor geolocation testbed are included. Finally, we close this paper with a short conclusion.

2. Wireless Geolocation Methods and Metrics

Most of the geolocation system architectures and methods developed for cellular systems are applicable for indoor geolocation systems although special considerations are needed for indoor radio channels. The most widely used wireless geolocation metrics include Angel of Arrival (AOA), Time of Arrival (TOA), Time Differences of Arrival (TDOA), Received Signal Strength (RSS) and Received Signal Phase. In this section we present an overview of overall system architectures and basic concepts of geolocation metrics as well as corresponding geolocation methods. The possibility of using these methods in indoor environment is also considered.

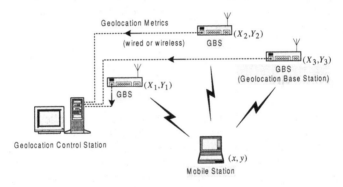

Fig. 1. Overall architecture of indoor geolocation system.

2.1 Overall System Architecture

Similar to the cellular geolocation system, the architecture of indoor geolocation systems can be roughly grouped into two main categories: mobile-based architecture and network-based architecture. Most of the indoor geolocation applications proposed to date have been focused on network-based system architecture as shown in Fig. 1 [4][5]. The geolocation base stations (GBS) extract location metrics from the radio signals transmitted by the mobile station and relay the information to a geolocation control station (GCS). The connection between GBS and GCS can be either wired or wireless. Then the position of the mobile station is estimated, displayed and tracked at the GCS. With the mobile-based system architecture, the mobile station estimates self-position by measuring received radio signals from multiple fixed GBS. Compared to mobile-based architecture, the network-based system has the advantage that the mobile station can be implemented as a simple-structured transceiver with small size and low power consumption that can be easily carried by people or attached to valuable equipments as a tag.

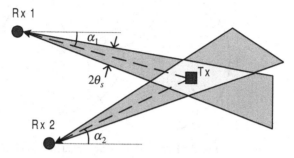

Fig. 2. Angel of Arrival geolocation method.

2.2 Angle of Arrival

The AOA geolocation method uses simple triangulation to locate the transmitter as shown in Fig. 2. The receiver measures the direction of received signals (i.e. angel of arrival) from the target transmitter using directional antennas or antenna arrays. If the accuracy of the direction measurement is $\pm\theta_s$, AOA measurement at the receiver will restrict the transmitter position around the line-of-sight (LOS) signal path with an angular spread of $2\theta_s$. AOA measurements at two receivers will provide a position fix as illustrated in Fig. 2. We can clearly observe that given the accuracy of AOA measurement, the accuracy of the position estimation depends on the transmitter position with respect to the receivers. When the transmitter lies between the two receivers, AOA measurements will not be able to provide a position fix. As a result, more than two receivers are normally needed to improve the location accuracy. For macro-cellular environment where the primary scatters are located around the transmitter and far away from the receivers, AOA method can provide acceptable location accuracy [6]. But dramatically large location errors will occur if the LOS signal path is blocked and the AOA of a reflected or a scattered signal component is used for estimation. In indoor environment, the LOS signal path is usually blocked by surrounding objects or walls. Thus AOA method will not be usable as the only metric for indoor geolocation system.

2.3 Time of Arrival and Time Difference of Arrival

The TOA method is based on estimating the propagation time of the signals (i.e. TOA) from a transmitter to multiple receivers. Several different methods can be used to obtain TOA or TDOA estimates, including pulse ranging [9][10], phase ranging [9] and spread-spectrum techniques [4][11].

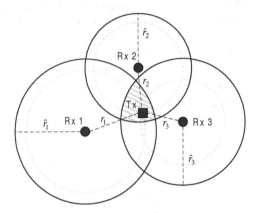

Fig. 3. Time of Arrival geolocation method.

Once TOA is measured, the distance between the transmitter and receiver can be simply determined since the propagation speed of the radio signal is approximately the speed of light $c = 3 \times 10^8$ m s^{-1}. The estimated distance at the receiver will geometrically define a circle, centered at the receiver, of possible transmitter positions. TOA measurements at three receivers will provide a position fix and given receiver coordinates and distances from the transmitter to receivers, the transmitter coordinates can be easily calculated. Due to multipath propagation, no-line-of-sight (NLOS) signal path and other impairments, the TOA-based distance estimates are always larger than the true distance between the transmitter and the receiver as illustrated in Fig. 2 where $\hat{r}_1$, $\hat{r}_2$ and $\hat{r}_3$ are the estimated distances and r_1, r_2 and r_3 are the true distances. Three TOA measurements determine a region of possible transmitter position as shown in Fig. 2. A nonlinear least square (NL-LS) method is usually used to obtain the best estimation iteratively by minimizing the estimation errors [6][4]:

$$e_i(x, y) = \hat{r}_i - \sqrt{(X_i - x)^2 + (Y_i - y)} \qquad (1)$$

for $i = 1, 2, ..., N$ where (X_i, Y_i) are receiver coordinates and (x, y) is transmitter coordinates. Sometimes the transmission time t_0 of the signals at the transmitter is also taken into account as the third variable:

$$e_i(x, y, t_0) = c(t_i - t_0) - \sqrt{(X_i - x)^2 + (Y_i - y)} \qquad (2)$$

where t_i is the receiving time of the signal at the i-th receiver. A constrained NL-LS algorithm is also available which makes use of the fact that TOA-based distance estimates are always larger than the true distance [8]. The same as in AOA method, more than three TOA measurements are needed to improve the accuracy of position estimation.

Instead of using TOA measurements, time difference measurements can also be employed to locate the receiver position. A constant time difference of arrival (TDOA) for two receivers defines a hyperbola, with foci at the receivers, on which the transmitter must be located. Three or more TDOA measurements provide a position fix at the intersection of hyperbolas. NL-LS method can also be used to obtain the best estimation of the transmitter position by minimizing the estimation error:

$$e_{i,j}(x,y) = c\hat{\tau}_{i,j} - \left[\sqrt{(X_i - x)^2 + (Y_i - y)^2} - \sqrt{(X_j - x)^2 + (Y_j - y)^2} \right] \tag{3}$$

for $i, j = 1, 2, ..., N$ where $\hat{\tau}_{i,j}$ is the TDOA measurement of ith and jth receivers.

There are some other methods to solve the hyperbolic position estimation problem as proposed in [12], [13], [14] and [15]. Compared to TOA method, the main advantage of TDOA method is that it does not require the knowledge of the transmit time from the transmitter while TOA method requires. As a result, strict time synchronization between transmitter and receivers is not required. However, TDOA method requires time synchronization among all the receivers.

2.4 Received Signal Strength

If the power transmitted by mobile terminal is known, measuring received signal strength (RSS) at receiver will provide the distance between the transmitter and the receiver using a known mathematical model for radio signal path loss with distances. The same as in the TOA method, the measured distance will determine a circle, centered at the receiver, on which the mobile transmitter must lie. Three RSS measurements will provide a position fix for the mobile. Due to shadow fading effects, RSS method results in large range estimation errors. The accuracy of this method can be improved by utilizing pre-measured received signal strength contour centered at the receiver [16]. A fuzzy logic algorithm was shown in [17] to be able to significantly improve the location accuracy.

2.5 Received Signal Phase

Signal phase is another possible geolocation metric. It is well known that with the aid of reference receivers to measure the carrier phase, differential GPS (DGPS) can improve the location accuracy from about 20m to within 1m compared to the standard GPS, which only uses pseudorange measurements [7]. One problem associated with the phase measurements lies in the ambiguity resulted from the periodicity (with period 2π) of the signal phase while the standard pseudorange measurements are unambiguous. Consequently, in the DGPS, the ambiguous carrier phase measurement is used in fine-tuning the pseudorange measurement. A complementary Kalman filter is used to combine the low noise ambiguous carrier phase measurements and the unambiguous but noisier pseudorange measurements [7]. For indoor geolocation system, it is possible to use the Received Signal Phase method together with TOA/TDOA or RSS method to fine-tune the location estimate. However unlike the application scenario of DGPS where LOS signal path is always observed, the multipath and no-line-of-sight condition of the indoor radio channel causes more errors in the phase measurements.

3. Example Wireless Indoor Geolocation Systems

Commercial indoor geolocation products have already appeared in the market. In this section, we present a technical overview of two example systems, Pinpoint Local Positioning System and Paltrack indoor geolocation system. Both companies claim that the indoor geolocation systems they developed can provide adequate location accuracy and location services in indoor environment.

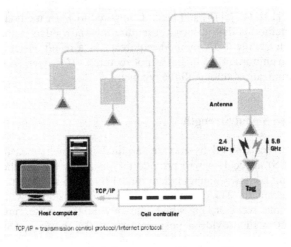

Fig. 4. PinPoint system architecture [4].

3.1 PinPoint Local Positioning System [4]

The system architecture of the PinPoint local position system is shown in Fig. 4. The PinPoint system uses simple-structured tags that can be attached to valuable assets or personnel badges. Indoor areas are divided into cells while each cell being served by a Cell Controller. The Cell Controller is connected to at most 16 antennas located at known positions. To locate tag position, Cell Controller transmits 2.4 GHz spread spectrum signal through different antennas in TDD mode. Once receiving signals from the Cell Controller antenna network, tags simply change the frequency of the received signal to 5.8 GHz and transmit back to the Cell Controller with tag ID information phase-modulated onto the signal. The distance between tag and antenna is determined by measuring round trip time of flight. With the measured distances from tag to antennas, the tag position can be obtained in the same way as in TOA method. A Host Computer is connected to Cell Controller through TCP/IP network to manage the location information of the tags. Since the Cell Controller generates the signal and measures round trip time of flight, there is no need to synchronize the clocks of tags and antennas.

The multipath effect is one of the limiting factors for indoor geolocation. Without multipath signal components, the time of arrival (TOA) could be easily determined from the triangular auto-correlation function of the spread spectrum signal. The triangular auto-correlation peak is two chips' (clock periods) wide at its base, and the time to rise from the noise floor to the peak is one chip. If the chipping rate were 1 MHz, it would take 1000 ns to rise from the noise floor to the peak, providing a "ruler" with a thousand 30-cm increments. A 40-MHz chipping rate was chosen for PinPoint system, providing a ruler of 25 ns that provides real-world increments of about 3.8 meters. Because of regulatory restrictions in the 2.44 and 5.78 GHz bands, faster chipping rates are not easy to achieve, and signal-processing techniques must be used to further improve the accuracy. Also to minimize the multipath effect, different frequency bands are used for uplink and downlink communication to avoid interference between the channels.

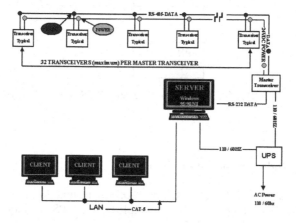

Fig. 5. Paltrack system architecture [5].

3.2 PalTrack Indoor Geolocation System [5]

The infrastructure of Paltrack indoor geolocation system, developed by Sovereign Technologies Corp., consists of tags, antennas, cell controllers and administrative software Server system as shown in Fig. 5. The PalTrack system utilizes a network structure that resides on an RS-485 node platform. A network of transceivers is located at known positions within the serving area while the transmitter tags are attached to assets. The tag transmitters transmit unique identification code at 418 MHz frequency band to a network of transceivers when on motion or at predefined time intervals. Transceivers estimates the tag location by measuring received signal strength (RSS) and utilizing a robust RSS-based algorithm patented by Sovereign Technologies Corp. The Master Transceiver collects measured information from the transceivers and relays it to a PC-based Server system. The accuracy for PalTrack is

0.6 to 2.4 m. The key component of the PalTrack system is the RSS-based geolocation algorithm.

4. An Indoor Geolocation Demonstrator

As we mentioned earlier, the indoor radio channel is very different from that of GPS systems or cellular system. To study the performance of various indoor geolocation techniques, a software indoor geolocation demonstrator is being developed at CWINS, WPI. Some general descriptions of the demonstrator as well as some simulation results will be presented in this section.

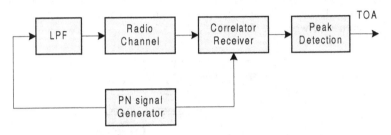

Fig. 6. Simulation of baseband DSSS geolocation system.

4.1 DSSS Indoor Geolocation System

The direct-sequence spread spectrum (DSSS) technique has been used in ranging systems for many years and is the principle behind GPS techniques. A block diagram of the simulated baseband DSSS geolocation system is shown in Fig. 6. For convenience, the PN (pseudo-random noise) signal generator is used in both transmitter and receiver with the assumption of time synchronization between geolocation transmitter and receiver. The lowpass filter (LPF) is used to take into account the band-limitation condition of the realistic radio transceiver systems. The autocorrelation characteristics of the PN sequence are fundamental to the distance (or TOA) estimation. A 31-chip Gold sequence was used with the chipping interval $T_c = 25\,\text{ns}$ and the sampling period $T_s = 5\,\text{ns}$ which provides a width of 50ns at the base of the triangular autocorrelation function of PN signal and an accuracy of 5ns in the TOA measurements. While the multipath radio channel spreads the transmitted signal, geolocation receivers are only interested in detecting the DLOS path, i.e. determining the arrival time of the first peak in the output autocorrelation signal [1]. With TOA measurements of signals from multiple reference transmitters, the position estimate of the receiver can be obtained iteratively using nonlinear least square algorithm.

4.2 Channel Measurement and Ray-Tracing

The channel profiles used in this simulation can be obtained in two ways. One is using a frequency domain measurement system described in [18]. The centerpiece in this system is a network analyzer that sweeps the channel from 900- 1100 MHz. The output signal is first amplified with an amplifier and then connected to the transmitter antenna through a long cable. The receiver antenna passes the signal through a chain of low noise amplifiers that are connected to input port of the network analyzer. The network analyzer records the frequency magnitude and phase responses of the channel. Fig. 7 shows an example of measured radio channel frequency response. For the geolocation simulation, we need to measure the actual frequency-domain channel response between the points of each reference transmitter and receiver. Then the channel impulse responses between transmitter and receiver, which we refer to as a channel time profile or simply a profile, can be obtained by taking the Fourier transform to the measured frequency channel response.

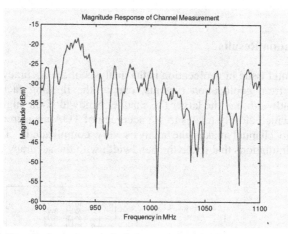

Fig. 7. Magnitude response of a frequency domain channel measurement.

Another way to obtain the channel profiles is using ray-tracing channel modeling method. Radio signals with frequencies larger than 300 MHz have extremely small wavelengths compared to the dimensions of building features so that electromagnetic waves can be treated simply as rays [19]. This is the principle behind ray-tracing method for radio channel modeling. In our simulation, we used CWINS 2D ray-tracing software to obtain the channel profiles, as shown in Fig. 8, between each reference transmitter and the receiver. Then the channel profiles can be directly used in geolocation simulations to obtain TOA and distance measurements.

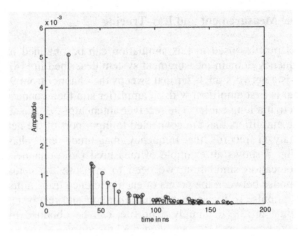

Fig. 8. Channel profiles obtained using 2-D ray-tracing software.

4.3 Simulation Results

A fundamental issue in geolocation is the analysis of the accuracy of positioning. For spread spectrum geolocation systems, one of the limiting factors is the available channel bandwidth, i.e. the larger the channel bandwidth, the higher the measurement resolution which closely relates to the accuracy of TOA measurement. The multipath indoor radio channel makes the analysis very complicated. This section presents results of simulations that relate the bandwidth with the accuracy of positioning.

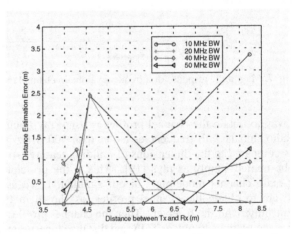

Fig. 9. Effects of channel bandwidth in ranging errors (with measurement-based method) [20].

Fig. 9 shows results obtained from channel measurement-based simulations. Table 1 shows the ranging errors between the transmitters and the receiver for different signal bandwidth obtained from both measurement based and ray-tracing based simulations. In Fig. 9 we observe that ranging errors are less than 3.5 m in all the cases. But to achieve less than 1.5 m accuracy, a bandwidth of larger than 20 MHz is needed.

Table 1. Mean ranging errors using measurement based and ray-tracing based methods [20][21].

Bandwidth (MHz)	Measurement based ranging error (m)	Ray-tracing based ranging error (m)
10	2.18	1.48
20	1.10	0.89
40	1.22	0.55
50	1.07	0.32

By using nonlinear least square algorithm to estimate receiver position from TOA measurements, we can employ more than two TOA measurements. Table 2 shows mean of the location estimation errors when different numbers of TOA measurements are used in the geolocation algorithm. Both measurement-based and ray-tracing based simulations show consistent results. But the ray-tracing based method is more convenient since for the measurement-based simulations, channel responses between all the transmitters and the receiver have to be measured.

Table 2. Mean location estimation errors with different numbers of TOA measurements [21].

Number of TOAs used in geolocation	Measurement based location error (m)	Ray-tracing based location error (m)
3	1.25	0.81
4	0.69	0.52
5	1.05	0.31
6	1.05	0.49
7	1.21	0.39

5. Conclusions

In this paper, we briefly reviewed various geolocation metrics and discussed usability of these metrics in indoor environment. Due to the serious multipath and no-line-of-sight propagation condition of the indoor radio channel, TOA/TDOA and Received Signal Strength methods are more appropriate than AOA method. Technical overview of two example wireless indoor geolocation products is then presented. Simulation results obtained from a software indoor geolocation testbed show that available channel bandwidth plays an important role in the accuracy of indoor geolocation systems. It was shown that ray-tracing based method provides consistent results with that of measurement-based method and the ray-tracing based method is more convenient than the frequency-domain measurement method.

Acknowledgement

The authors would like to express their appreciation to TEKES, Nokia, and Finnish Airforce for supporting most parts of this project. We are thankful to two MQP groups at CWINS, WPI for their contributions to this work: K.H. Shah, C.M. Kelly and D.S. Hastings for measurement-based simulations and S. Dasmah, C.D. Le and T.Q. Nguyen for ray-tracing based simulations. We also thank Dr. Jacques Beneat and Dr. Prashant Krishnamurthy, our colleagues at CWINS, for fruitful discussions and a variety of help.

References

[1] K. Pahlavan, P. Krishnamurthy and J. Beneat, "Wideband radio propagation modeling for indoor geolocation applications", IEEE Comm. Magazine, pp. 60-65, April 1998.

[2] P. Krishnamurthy, K. Pahlavan, J. Beneat, "Radio propagation modeling for indoor geolocation applications", Proceedings of IEEE PIMRC'98, September 1998.

[3] Jay Werb and Colin Lanzl, "Designing a positioning system for finding things and people indoors", IEEE Spectrum, vol. 35, No. 9, Sep. 1998.

[4] PinPoint Local Positioning System, http://www.pinpointco.com/.

[5] PalTrack Tracking Systems, http://www.sovtechcorp.com/.

[6] J. Caffery, Jr. and G.L. Stuber, "Subscriber Location in CDMA Cellular Networks", IEEE Trans. Veh. Technol., vol. 47, No. 2, May 1998.

[7] E.D. Kaplan, Understanding GPS: Principles and Applications, Artech House Publishers, 1996.

[8] G. Morley and W. Grover, "Improved location estimation with pulse-ranging in presence of shadowing and multipath excess-delay effects", Electronic Letter, vol. 31, pp 1609-1610, Aug., 1995.

[9] G. Turin, W. Jewell and T. Johnston, "Simulation of urban vehicle-monitoring systems", IEEE Trans. Veh. Technol., vol. VT-21, pp. 9-16, Feb. 1972.

[10] H. Hashemi, "Pulse ranging radiolocation technique and its application to channel assignment in digital cellular radio", Proc. IEEE VTC'91, pp. 675-680, 1991.

[11] P. Goud, A. Sesay and M. Fattouche, "A spread spectrum radiolocation technique and its application to cellular radio", Proc. IEEE Pacific Rim Conf. Comm., Comp. and Signal processing, 1991, pp. 661-664.

[12] W.H. Foy, "Position-location solutions by Taylor-series estimation", IEEE Trans. Aerospace and Electronic Systems, vol. AES-12, pp. 187-194, Mar. 1976.

[13] D.J. Torrieri, "Statistical theory of passive location system", IEEE Trans. Aerospace and Electric Systems, vol. AES-20, No. 2, Mar. 1992.

[14] J.S. Abel and J.O.Smith, "A divide-and-conquer approach to least-squares estimation", IEEE. Trans. Aerospace and Electric Systems, vol. 26, pp. 423-427, Mar. 1990.

[15] Y.T. Chan and K.C. Ho, "A simple and efficient estimator for hyperbolic location", IEEE Trans. Signal Processing, vol. 42, No. 8, pp. 1905-1915, Aug. 1994.

[16] W. Figel, N. Shepherd and W. Trammell, "Vehicle location by a signal attenuation method", IEEE Trans. Vehicular Technology, vol. VT-18, pp. 105-110, Nov. 1969.

[17] Han-Lee Song, "Automatic Vehicle Location in Cellular Communications Systems", IEEE Trans. Vehicular Technology, vol. 43, No. 4, pp. 902-908, Nov. 1994.

[18] S.J. Howard and K. Pahlavan, "Measurement and analysis of the indoor radio channel in the frequency domain", IEEE Trans. Instr. Meas. No. 39, pp. 751-755, 1990.

[19] K. Pahlavan and A. Levesque, Wireless Information Networks, John Wiley and Sons, New York, 1995.

[20] K.H. Shah, C.M. Kelly and D.S. Hastings, MQP Project Report: Wireless Indoor Geolocation System, CWINS, Worcester Polytechnic Institute, May 1999.

[21] S. Dasmah, C.D. Le and T.Q. Nguyen, MQP Project Report: Simulation Platform for Performance Evaluation of Indoor Geolocation, CWINS, Worcester Polytechnic Institute, January 2000.

Priority Based Multiple Access for Service Differentiation in Wireless Ad-Hoc Networks

Yu Wang[1] and Brahim Bensaou[2]

[1] Department of Electrical Engineering, National University of Singapore,
10 Kent Ridge Crescent, Singapore 119260
engp8843@nus.edu.sg

[2] Centre for Wireless Communications, National University of Singapore,
20 Science Park Road, #02-34/37, Singapore 117674
brahim@cwc.nus.edu.sg

Abstract. This article describes the Priority Based Multiple Access (PriMA) protocol, a new medium access control (MAC) protocol for single-channel ad-hoc networks. Unlike previously proposed protocols, PriMA takes into account the QoS requirements of the packets queued in stations to provide each station with a priority-based access to the channel. The direct support of PriMA for ad-hoc routing is that when some stations act as hubs in the routing structure and route packets for other stations besides their own, they can have high priorities and obtain larger share of bandwidth. Simulation results show the potential benefits that PriMA brings about to ad-hoc networks, and confirm PriMA as an initial step towards QoS provision in ad-hoc networks.

1 Introduction

An ad-hoc network is a dynamic multi-hop wireless network that is established by a group of mobile stations without the aid of any pre-existing network infrastructure or centralized administration. It can be installed quickly in emergency or some other special situations and is self-configurable, which makes it very attractive in many applications. Some applications include remote sensing (e.g. earthquakes, fire, environmental data gathering), robotic communication (e.g. cleaning, firefighting, patrolling, oceanic fishing, farming), Internet access and LEO constellations. Internet Engineering Task Force (IETF) has set up a workgroup named Mobile Ad-hoc Networks (MANET) [1] to develop and evolve MANET routing specifications and introduce them to the Internet Standards track. Recent discussions in MANET's mailing list on MANET's application scenarios can give an idea of the fields when MANETs can be deployed (e.g. an ad-hoc network of New York taxi cabs).

An efficient medium access control (MAC) protocol through which mobile stations can share a common broadcast channel is essential in an ad-hoc network because the medium or channel is a scarce resource. Due to the limited transmission range of mobile stations, multiple transmitters within range of the

C.G. Omidyar (Ed.): MWCN 2000, LNCS 1818, pp. 14–30, 2000.

same receiver may not know one another's transmissions, and thus in effect "hidden" from one another. When these transmitters transmit to the same receiver at around the same time, they do not realize that their transmissions collide at the receiver. This is the so-called "hidden terminal" problem [2] which is known to degrade throughput significantly. Due to their multi-hop characteristics, ad-hoc networks suffer much more from the hidden terminal problem than wireless LANs do.

To address the hidden terminal problem, various distributed MAC protocols were proposed in the literature. MACA (Multiple Access Collision Avoidance) protocol [3] proposed exchange of short Request-to-Send (RTS) and Clear-to-Send (CTS) packets between a pair of sender and receiver before actual data packet transmission and formed the basis for several other more sophisticated schemes. Among them, FAMA-NCS (Floor Acquisition Multiple Access with Non-persistent Carrier Sensing) [4] is immune to the hidden terminal problem and can achieve good throughput in ad-hoc networks. IEEE 802.11 committee also proposed a MAC protocol called Distributed Foundation Wireless Medium Access Control (DFWMAC) for wireless ad-hoc LANs [5], which in essence is a variant of CSMA/CA protocols. DFWMAC provides basic and RTS/CTS access method. Chhaya and Gupta [6] have shown that the performance of RTS/CTS access method degrades much slower than the basic access method when the number of hidden terminals is large, or the offered load is significantly larger than the channel capacity; therefore, the RTS/CTS method is more robust to fluctuations in parameter values which are common in ad-hoc networks. In DFW-MAC RTS/CTS access method, 4-way dialog including RTS-CTS-DATA-ACK is used to combat the hidden terminal problem. DFWMAC, however, still cannot prevent data packets from colliding with control packets (RTSs and CTSs) and other data packets, instead, it uses a sophisticated modified binary exponential backoff scheme to quickly resolve collisions and thus increase throughput.

These MAC protocols perform well in solving hidden terminal problem, however, none of them takes any step towards providing packet level QoS parameters such as packet loss ratio, packet delay, etc. They can support only best effort delivery service, thus limit the applications of ad-hoc networks. Recently, there has been a surge in modifying and extending equality-based medium access schemes to support priority-based access when the packets queued at stations have different QoS requirements. Although this research effort does not compete with the effort deployed in the Wireless LAN (infrastructure based), some researchers have been working on QoS support for distributed wireless networks [7, 8]. In [7], stations with real-time packets in the queue would jam the channel with Black Bursts (BBs) whose length is proportional to the delay incurred. The station that sends the longest BBs wins access to the channel and can transmit its packet thereafter. However, this approach fails when hidden terminals exist as those hidden terminals may have experienced the same delay and each BB contention period is not guaranteed to produce a unique winner, thus real-time data packets will still suffer from collisions. In [8], GAMA (Group Allocation Multiple Access) protocol was proposed for scheduling real-time and datagram traffic in a single-hop wireless ad-hoc LANs. GAMA includes contention period during which stations can send request to join transmission group and contention-free period during which stations in transmission group take turns to transmit packets. This approach does not work well if hidden terminal exists. If hidden terminals do not join the transmission group that it may interfere with, GAMA cannot ensure data packets to be free from collision. If hidden terminals do join the transmission group to avoid collision, following the same logic, then all the other stations in the network have to join the same transmission group one by one. It is very difficult to maintain the global group due to the dynamic nature of ad-hoc network. In addition, we cannot benefit from spatial reuse. Therefore, these protocols that were originally proposed for wireless LANs are not directly applicable to multi-hop ad-hoc networks. We believe that a suitable MAC protocol for ad-hoc networks should address both hidden terminal problem and QoS issues at the same time.

Another motivation for priority-based access is to provide better support for ad-hoc routing. On one hand, those protocols originally proposed for wireless LANs normally do not take routing into account as they expect a wireless access point that can reach all other stations and relay packets for them to be deployed. Therefore they are more fit for networks with infrastructure than ad-hoc networks where routing is another major issue. On the other hand, unlike in conventional wired networks, any host that act as a router in ad-hoc networks normally has only one network interface, there are no separate links for them to route packets or exchange routing information. Especially when some stations in an ad-hoc network behave as cluster heads [9] or belong to the core of the routing structure [10], more traffic will transit through them. Obviously, they need higher priorities in accessing the channel to route packets for others in addition to their own.

In this paper, we propose a new protocol named Priority Based Multiple Access (PriMA) to support differentiated access priorities to the channel. It implements MAC-level acknowledgment as in DFWMAC while adopting the collision-free data transmission characteristics of FAMA-NCS. More importantly, it implements a novel distributed scheduling algorithm which gives stations a dynamic-priority based access to the channel by taking into account both the packet delay requirement and the packet loss ratio incurred by the ongoing session. The remainder of the paper is organized as follows. Section 2 describes PriMA protocol in detail. Section 3 compares by simulation the performance of FAMA-NCS, IEEE 802.11 DFWMAC and PriMA. Section 4 concludes this paper.

2 PriMA Protocol

2.1 Overview

PriMA protocol requires a station that wishes to send a data packet to acquire the channel before transmitting the packet. The channel is acquired by establishing an RTS-CTS dialog between the sender and the receiver. Although multiple control packets may collide, data packets are always sent free from collisions. This is achieved by the enforced requirements of size relationship between RTS and CTS as well as different periods a station should wait after receiving a packet or sensing the channel busy. Fullmer et al. [11] have given detailed description about this. They showed that as long as the length of CTS packet is sufficiently longer than that of the RTS packet, CTS can act as a jamming signal to prevent other stations from transmission. Stations that hear the channel busy should wait long enough for the possible ongoing data transmission to go on unobstructed. PriMA's trivial extension to this is to reply an acknowledgment packet after successful reception of a data packet. This is also the common case in some other MAC protocols like DFWMAC. The extension of MAC-level acknowledgment still ensures data transmission free from collisions.

The most important feature of PriMA is that the access to the channel is based on priorities of the packets queued in stations. This is achieved by three timers calculated according to the QoS requirements of the data packets in each station. The first is the *access timer* that a station should wait after the channel becomes idle before transmitting an RTS packet. The access time is similar to DIFS (Distributed InterFrame Space) in DFWMAC and the silent period of channel in FAMA-NCS. However, unlike the fixed length of DIFS and the random length of the silent period in FAMA-NCS, the access time is dynamically adjusted based on the QoS requirements. The second is the *delay timer* carried in every RTS packet. It indicates how long the intended receiver can wait before replying with a CTS packet. Therefore, an earlier sent RTS packet may be preempted by a later sent RTS packet which indicates a higher priority data packet transmission request. The third is the *backoff time* that a station should wait before retransmission when collision occurs. The backoff time is uniformly distributed, however the upper bounds of the distribution vary among stations. Stations that hold higher priority packets have lower upper bounds. Therefore, they can statistically recover earlier than other stations and bid for the channel again. For stations queued with normal data packets, we just set a large upper bound comparable to that of DFWMAC. Therefore, our backoff scheme is not necessarily more prone to collapse than binary exponential backoff with upper bound.

2.2 Description of PriMA

To simplify our description of PriMA, the processing time and transmit-to-receive turnaround time are ignored. We say a station "detects" collisions if it senses the channel busy without being able to receive any intelligible packet. Following, we define some of the notations used in this section:

rtPacket : Data packet with QoS parameters specified
nrtPacket : Data packet without any QoS parameters specified
T_d : Maximum one-hop channel propagation delay

T_{type} : Time to transmit a packet of type *type* where *type* can be either RTS, CTS, DATA or ACK

T_{delay} : Allowed delay time to reply to RTS, carried in RTS packet

T_{left} : Time left for a *rtPacket* to be delivered; when T_{left} is less than a threshold, the corresponding *rtPacket* will be dropped.

T_{max} : Maximum time to complete one successful RTS-CTS-DATA-ACK transmission

T_{access} : Time required to sense the channel idle before transmitting an RTS packet

T_{defer} : The longest time a station should defer access when "detecting" collisions before entering backoff. It equals $T_{data} + 3 * T_d$.

T_{unit} : Time used as a factor to map packet delay requirement to backoff timer

N_{lost} : Number of packets dropped during the session

N_{sent} : Number of packets sent during a session

PLR : Packet Loss Ratio (QoS requirement).

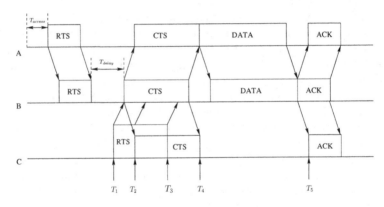

Fig. 1. PriMA illustration

Figure 1 shows how PriMA operates. In this figure, stations A and C are hidden from each other and station B is a neighbor of both A and C. As in all other collision avoidance multiple access techniques PriMA uses the RTS-CTS combination to implement collision avoidance, besides to ensure the problem of hidden terminals is addressed adequately, PriMA imposes FAMA's conditions on the sizes of the RTS and CTS packet [11]:

- $T_{rts} > 2 * T_d$
- $T_{cts} > T_{rts} + 2 * T_d$

and stations that sense the channel busy should wait for T_{defer} to let the possible ongoing data transmission to finish unobstructed. FAMA's rational in imposing these conditions, is that although C is hidden from A, and thus C did not hear

A's RTS, if C sends an RTS packet that at most collides with the CTS reply from B to A, the condition ensures that C hears the trailer of the CTS and thus C would abort any transmission and enters the backoff procedure. This ensures that the data packet from A to B is transmitted collision free. Note that it is not necessary to adopt these restrictions (enforced in FAMA) in PriMA to provide soft QoS. Nevertheless, we believe that it does not make sense to address the problem of QoS (be it soft QoS) without addressing first the hidden terminal problem, especially when the offered traffic load can be very high. PriMA's approach to providing QoS can also be applied to IEEE 802.11's DFWMAC.

In PriMA, any station that has a packet to send should wait until it senses the channel idle for a certain amount of time called *access timer* before transmitting an RTS request. The access time calculation, shown in Algorithm 1, is based on the required QoS and the perceived QoS.

Algorithm 1 Access timer calculation

1: **if** (nrtPacket or rtPacket with $T_{left} > 1$ sec) **then**
2:     $T_{access} = 2 * T_d$
3: **else**
4:     **if** $(T_{left} > 0)$ **then**
5:        $T_{access} = T_{left} - (N_{lost} - PLR * N_{sent}) * T_{max}$
6:        **if** $(T_{access} < 0)$ **then**
7:           $T_{access} = 0$
8:        **else**
9:           $T_{access} = 2 * T_d / |\log_2 (T_{access}/2)|$
10:        **end if**
11:     **end if**
12: **end if**

Note that in order to reduce the waste of bandwidth due to this access time, time-sensitive packets which have a time to live of more than one second, compete at the same priority level as the non time sensitive packets. These packets would gain higher priorities when they become more urgent. Note that initially, at the packet's first attempt, T_{left} is initialized to a value proportional to the maximum tolerable delay of the packet. Also, throughout the algorithm, T_{left} decreases with the ticks of the clock in the same proportion. The calculation in line 5 of the algorithm shows that the more has a station suffered excess droppped packets $(N_{lost} - PLR * N_{sent})$, the shorter T_{access} will be. This excerpt of pseudo-code clearly shows that when two stations attempt to access the channel simultaneously T_{access} will differentiate their attempts according to their packets' QoS requirement.

When a station succeeds in its access to the channel, it sets a "delay" field in its RTS packet and sends it out. The value of the field is denoted by T_{delay}, which indicates how long the intended receiver can wait before replying with a CTS packet. T_{delay} is calculated as shown in Algorithm 2.

Algorithm 2 Delay timer calculation

1: **if** (nrtPacket) **then**
2:     $T_{delay} = 2 * T_{rts}$
3: **else**
4:     **if** $(T_{left}/T_{max} > 2)$ **then**
5:         $T_{delay} = 2 * T_{rts}$
6:     **else**
7:         $T_{delay} = (T_{left}/T_{max}) * T_{rts}$
8:     **end if**
9: **end if**

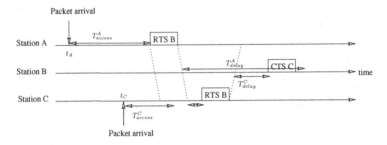

Fig. 2. Illustration of the preemption in PriMA

To illustrate the importance of this second timer, let us consider the example in Figure 2 where two stations, A and C, attempt to communicate with station B. We will neglect the propagation times in this example. C's data packet arrives at the MAC layer at time t_C later than A's data packet who arrives at t_A, however, C's delay target is more stringent than A's. Due to the lack of a global coordinator in the system and the lack of a global state, there is no way for C to succeed in sending an RTS packet to B before A does, since $t_C + T^C_{accces} > t_A + T^A_{access}$. A sets a value T^A_{delay} in its RTS packet which tells B the amount of time it can wait before replying with a CTS. After C hears the end of A's RTS (we will see later the case where A and C are hidden from each other), it will apply the same procedure it uses for the first access, however with only the remaining access timer. In other words, C will wait until it senses an idle channel for $T_{left} = t_C + T^C_{accces} - t_A - T^A_{access}$ starting at $t_A + T^A_{access} + T_{RTS}$. If this left time is smaller than T^A_{delay}, i.e., if C's access timer expires before B replies with a CTS to A than C can send to B an RTS that will preempt A's RTS. This way, although A succeeded in sending the RTS before C, this latter has the possibility to preempt A before it starts transmitting its data packet.

In the calculation of T_{delay} for a real time packet, only T_{left} is used. This is because T_{left} is a composite value affected by both packet delay and packet drop ratio. Thus we can keep the algorithm simple yet effective. This algorithm shows that the RTS packet for a data packet that is more delay sensitive will have shorter T_{delay} when the data packet is about to expire. The calculation will permit the packets to gain higher priorities after they are delayed for some time. Stations usually do not reply with CTS immediately after receiving an RTS packet unless required to do so, therefore a station may receive multiple RTS packets in a row. If the later arrived RTS packet requires shorter delay than that of the earlier one, the station can reply to it first. This makes it possible for delay sensitive data packets to preempt other data packets. In addition, T_{delay} does not exceed two times the RTS packet transmission time to minimize the protocol overhead.

It is inevitable that control packets may collide with each other, therefore those stations that detect collisions should also wait for T_{defer} and then back off for a random time $T_{backoff}$, which is calculated as shown in Algorithm 3, where $U(0, x)$ is a uniformly distributed random number in the interval $(0, x)$.

Algorithm 3 Backoff timer calculation

1: **if** (nrtPacket or rtPacket with $T_{left}/T_{unit} > Maxtimer$) **then**
2:     $T_{backoff} = U(0, Maxtimer) * 2T_d$
3: **else**
4:     $T_{backoff} = U(0, T_{left}/T_{unit}) * 2T_d$
5: **end if**

The value of *Maxtimer* is simply set to a large value, say 800, which is comparable to the maximum timer used in DFWMAC. For example, the IEEE 802.11 Wireless LAN standard specification for direct sequence spread spectrum (DSSS) uses a timer (or *Contention Window*, in 802.11's terminology) ranging from 31 to 1023. In PriMA, $T_{backoff}$ is based on a uniform distribution whose upper bound value varies according to the delay requirement of a data packet. This calculation will statistically give stations that have delay sensitive data packets shorter $T_{backoff}$. Thus, on average, these stations will end the backoff period earlier than other stations and bid for the channel again. To illustrate how Algorithm 3 contributes in differentiating the stations, let us return to the example of Figure 2. In the case where A and C are hidden from each other. C cannot hear the RTS packet of A and thus after exhausting its timer T_{access}^C, it would also send an RTS packet which would collide at B with A's RTS packet. After T_{delay}^A (respectively T_{delay}^C) station A (respectively station C) notices the collision by timeout, and thus they both apply the above backoff algorithm with their respective QoS targets. In this case, C has higher probability in bidding for the channel before A does, since C draws its uniformly distributed backoff time from a smaller interval than A does, and thus C has higher probability to terminate its backoff before A.

From the above descriptions, we can see that by dynamically adjusting three timers, T_{access}, T_{delay} and $T_{backoff}$, according to the data packets' QoS requirements, PriMA can give stations priority-based access to the channel. At the same time, various waiting time periods are carefully defined to prevent control packets from colliding with data packets.

3 Simulation Results

PriMA is a complex MAC protocol for analytical modelling because of the danymic nature of the different timers, thus in this paper the performance of PriMA is investigated by simulations. In our experiments, we investigate symmetrical networks where each station has N neighbors and is hidden from Q neighbors of any one of its neighbors, thus each station has the same spatial characteristics. Figure 3 shows two sample configurations for $N = 4$, $Q = 2$ and $Q = 3$. For example, for station A in figure 3(a), it has 4 neighbors including station B and is hidden from station B's 3 neighbors: C, D and E. As this graph can grow to infinitely large, we fold or collapse it so that the total number of stations in the network is limited while symmetry is maintained. The resulting graphs are shown in figure 4.

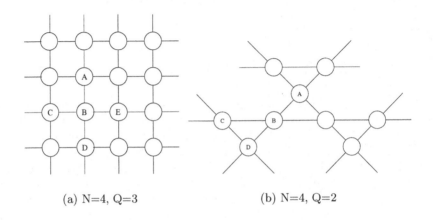

(a) N=4, Q=3 (b) N=4, Q=2

Fig. 3. Two sample configurations

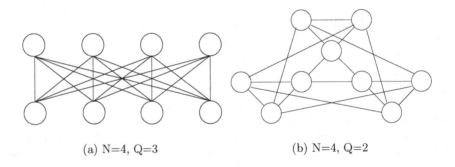

(a) N=4, Q=3 (b) N=4, Q=2

Fig. 4. Two sample configurations (collapsed)

We assume a 1Mbps ideal channel with zero preamble and processing overhead. We have performed different sets of simulations with OPNET Modeler/Radio and we compare our results to those provided by alternative protocols, notably FAMA-NCS and IEEE 802.11 DFWMAC[1]. Table 1 lists the parameters used in the simulation. As we ignore the extra time incurred by hardware and software, the different InterFrame Spaces (IFSs) in IEEE 802.11 are reduced accordingly and they are shown in table 2.

Protocol	RTS	CTS	DATA	ACK	backoff timer	backoff unit time
FAMA	20-byte	25-byte	500-byte	-	10	$160\mu sec$
IEEE 802.11	25-byte	20-byte	500-byte	20-byte	31-1023	$6\mu sec$
PriMA	20-byte	25-byte	500-byte	20-byte	800	$12\mu sec$

Table 1. Protocol configuration parameters

DIFS	SIFS	EIFS
$12\mu sec$	$0\mu sec$	1.3msec

Table 2. Extra configuration parameters for IEEE 802.11

In the first set of simulations, all stations generate Poisson traffic with the same mean rate and all require best-effort delivery service. The simulation measures the throughput of the protocol against the degree of the nodes. The results are shown in Figure 5. For comparative purpose, we also show the analytical results for Slotted-ALOHA with separate acknowledgement channel and Non-Persistent CSMA in these figures. The figures demonstrate that FAMA's performance degrades dramatically when the number of competing stations (including neighbors and hidden terminals) increases. This is due to the ineffectiveness of uniform backoff scheme with small timer in collision resolution used in FAMA. However, it still performs well in other situations because of its immunity to hidden terminal problem. IEEE 802.11 performs quite well in all the situations because of its modified binary exponential backoff scheme. Although PriMA's performance is not quite outstanding when the number of competing stations is small, however, PriMA does achieve a rather stable throughput and is comparable to IEEE 802.11 when the number of competing stations increases. This is not surprising. PriMA, by introducing the different timers as well as the MAC acknowledgments in fact sacrifices throughput to achieve different priorities among the stations (which we show later). In addition, large backoff timers will become more effective when the number of competing stations increases, therefore the throughput difference between PriMA and IEEE 802.11 is decreasing. In other

---

[1] We use its specification for Direct Sequence Spread Spectrum if applicable.

words, PriMA captures in fact the effectiveness of FAMA in solving the hidden terminal problem as well as 802.11's ability in sustaining throughput at high loads. It sacrifices however a little throughput by introducing the different timers in order to achieve differentiation between the stations as shown in the following.

In the second set of simulations, each station still generates Poisson traffic with the same mean rate. However, we set one station (router) to generate packets that require delay-bounded delivery, while the other stations (hosts) have no delay requirement. This, in essence, makes the packets from the router have higher priority than packets from the hosts.

This scenario may be applicable in a case when some stations form a group and choose one of them to act as a router. In other words, the router will handle a larger amount of traffic and thus needs some priority to access the channel. Simulation results are shown in figure 6 and 7 with different values of N and Q. The figure clearly shows that the router has higher throughput than other stations despite the fact that they all offer the same traffic load to the channel. Even when the number of competing stations increases, the router can still achieve much higher throughput than other normal stations. This differentiation cannot be achieved by FAMA-NCS and other non prioritized protocols such as IEEE 802.11 DFWMAC.

In the third set of simulations, we select two stations (high priority router and low priority router) to generate constant rate packets that require different packet delivery delay bounds. We investigate two performance metrics under different scenarios: namely, the packet loss ratio (due to expiration), and the average delay. To get comparative results, we set the packet interarrival times equal to the packet delay requirement, otherwise protocols like IEEE 802.11 will never keep up with the packet generation rate and eventually will drop nearly all the packets. Simulation results are shown in figure 8. The first three sub-figures (8(a), 8(b), 8(c)) show the packet loss ratio against the offered load. They show that when the offered load increases, the packet drop ratio for IEEE 802.11 increases significantly while PriMA can still maintain low packet drop ratio for the two routers. Figure 8(d) shows that the average packet delay achieved by the two types of routers in PriMA. The figure clearly shows that when the traffic load is very low, the packets from all the stations experience the same (insignificant) delay. However, when the traffic load increases, the protocol starts differentiating the stations according to their requirements.

4   Conclusion

PriMA is a new MAC protocol that is specifically designed for ad-hoc networks. It can achieve good throughput in ad-hoc networks where hidden terminal problem is common. In addition, every station has priority-based access to the channel, thus PriMA can provide elementary QoS support from the bottom up, making it a good choice for supporting higher layer protocols that require quality of service. Another benefit of PriMA is that it can provide better support for

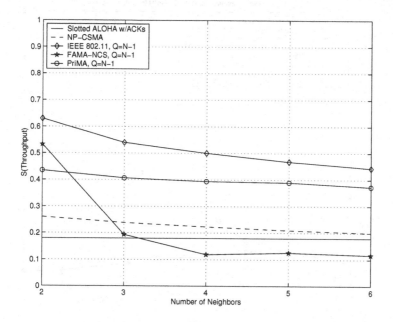

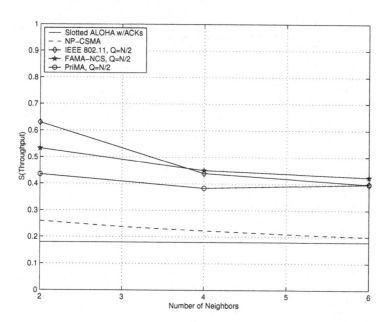

Fig. 5. Throughput versus node degree

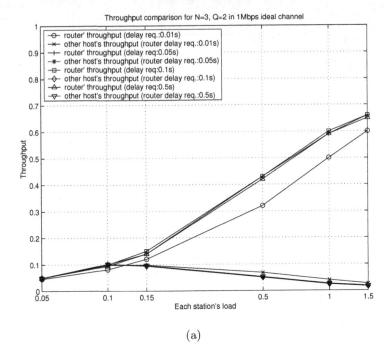

(a)

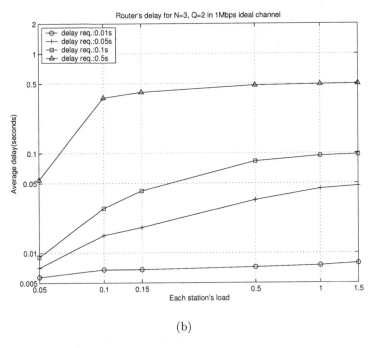

(b)

Fig. 6. One router configuration (N=3)

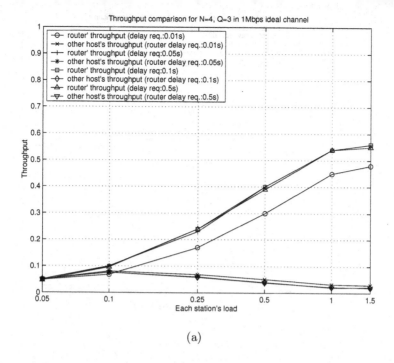

(a)

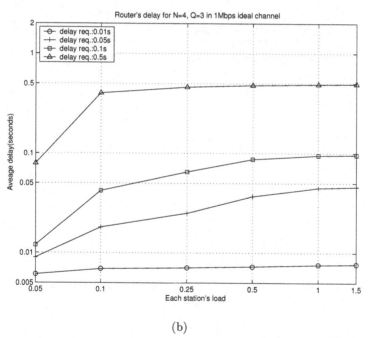

(b)

Fig. 7. One router configuration (N=4)

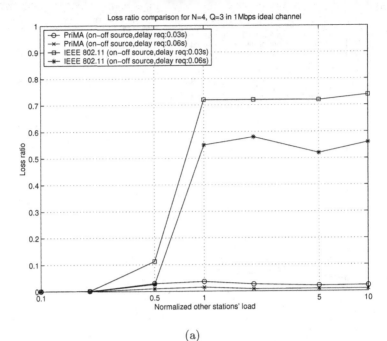

(a)

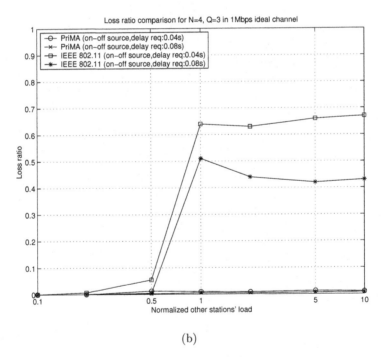

(b)

Fig. 8. QoS differentiation

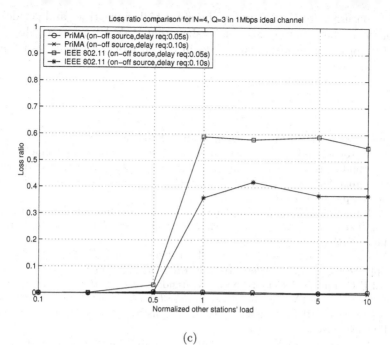

(c)

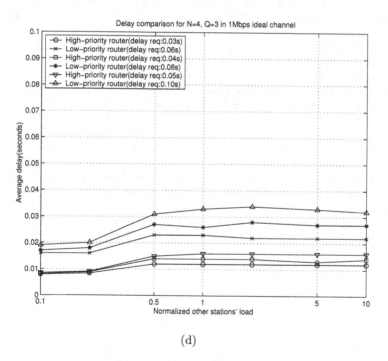

(d)

Fig. 8. QoS differentiation (con't)

ad-hoc routing as some packets can be given higher priorities and get delivered earlier than others. Simulation results show that PriMA is an especially promising MAC protocol for ad-hoc networks in presence of heterogeneous traffic and QoS requirements.

References

[1] http://www.ietf.org/html.charters/manet-charter.html.

[2] F. A. Tobagi and L. Kleinrock, "Packet Switching in Radio Channels: Part II - the Hidden Terminal Problem in Carrier Sense Multiple-access Modes and the Busy-tone Solution," in *IEEE Transactions on Communications, vol. COM-23, no. 12*, pp. 1417–1433, 1975.

[3] P. Karn, "MACA - a New Channel Access Method for Packet Radio," in *ARRL/CRRL Amateur Radio 9th Computer Networking Conference*, pp. 134–140, ARRL, 1990.

[4] J. Garcia-Luna-Aceves and C. L. Fullmer, "Performance of Floor Acquisition Multiple Access in Ad-Hoc Networks," in *Proceedings of 3rd IEEE ISCC*, 1998.

[5] IEEE Computer Society LAN MAN Standards Committee, ed., *IEEE Standard for Wireless LAN Medium Access Control (MAC) and Physical Layer (PHY) Specifications*. IEEE Std 802.11-1997, The Institute of Electrical and Electronics Engineers, New York, 1997.

[6] H. S. Chhaya and S. Gupta, "Throughput and Fairness Properties of Asynchronous Data Transfer Methods in the IEEE 802.11 MAC Protocol," in *6th International Conference on Personal, Indoor and Mobile Radio Communications*, 1995.

[7] J. L. Sobrinho and A. S. Krishnakumar, "Quality-of-Service in Ad Hoc Carrier Sense Multiple Access Wireless Networks," in *IEEE Journal on Selected Areas in Communications, Vol. 17, No. 8*, pp. 1353–1368, 1999.

[8] A. Muir and J. J. Garcia-Luna-Aceves, "Supporting real-time multimedia traffic in a wireless LAN," in *Proc. SPIE Multimedia Computing and Networking*, pp. 41–54, 1997.

[9] M. L. Jiang, J. Y. Li, and Y. C. Tay, "Cluster Based Routing Protocol(CBRP) Functional Specification." draft-ietf-manet-cbrp-spec-00.txt, Aug. 1998. Work in progress.

[10] R. Sivakumar, P. Sinha, and V. Bharghavan, "Core Extraction Distributed Ad hoc Routing (CEDAR) Specification." draft-ietf-manet-cedar-spec-00.txt, Oct. 1998. Work in progress.

[11] C. L. Fullmer and J. J. Garcia-Luna-Aceves, "Solutions to Hidden Terminal Problems in Wireless Networks," in *Proceedings of ACM SIGCOMM*, 1997.

Survivability Analysis of
Ad Hoc Wireless Network Architecture

Krishna Paul[1] , Romit RoyChoudhuri[2], Somprakash Bandyopadhyay[3]

[1] Cognizant Technology Solutions, Sector V, Saltlake
Calcutta 700 091 India
{PKrishna2@cal.cts-corp.com}

[2] Department of Computer Sc and Engg
Haldia Institute of Technology
Haldia, West Bengal, India

[3] PricewaterhouseCoopers, Saltlake Technology Center
Sector V, Calcutta 700 091, India
{somprakash.bandyopadhyay@in.pwcglobal.com}

Abstract. Mobile ad hoc wireless networks are generating novel interests in mobile computing. The dynamism in network topology has thrown up multifarious issues, requiring a fresh look into the aspects of system design and networking protocols. As a direct consequence of injecting mobility into a static network, the formal relationships between several governing parameters have undergone changes. In this paper we have assayed the behavior of the ad hoc network as a whole and analyzed trends in the inter-parameter dependencies, with the objective of addressing to the survivability issues. We have finally drawn out an operating region of survivability for mobile ad hoc wireless networks in terms of user declared specifications. Our own simulator has been operative through the work. We have derived our survivability constraints from several runs of the network simulator.

1. Introduction

An ad hoc network [1] can be envisioned as a collection of mobile routers, each equipped with a wireless transceiver, which are free to move about arbitrarily. The mobility of the routers and the variability of other connecting factors results in a network with a potentially rapid and unpredictable changing topology. These networks may or may not be connected with the infrastructure such as internet, but still be available for use by a group of wireless mobile hosts that operates without any base-station or any centralized control. Applications of ad hoc networks include military tactical communication, emergency relief operations, commercial and educational use in remote areas, etc. where the networking is mission-oriented and / or community-based.

There has been a growing interest in ad hoc networks in recent years [1,2,3,4,5]. The basic assumption in an ad-hoc network is that, two nodes willing to communicate may be outside the wireless transmission range of each other but still be able to

C.G. Omidyar (Ed.): MWCN 2000, LNCS 1818, pp. 31-46, 2000.
© Springer-Verlag Berlin Heidelberg 2000

communicate if other nodes in the network are willing to forward packets from them. However, the successful operation of an ad-hoc network will be hampered, if an intermediate node, participating in a communication between two nodes, moves out of range suddenly or switches itself off in between message transfer. The situation is worse, if there is no other path between those two nodes. An important problem associated with this is to find a stable path satisfying multiple constraints to ensure certain level of QoS guarantee during communication.

Lot of research has been done on ad hoc network routing protocols in order to solve the problem of routing packets.   However, there is no complete proposal available to assess the survivability issues in ad hoc network in order to provide a network specification to support effective communication in such a dynamic environment. Survivability analysis [6], in this context, can be defined as network specifications and management procedures to minimize the impact of system dynamics on the network services . For example, assume an area 1000 x 1000 sq.meter where 20 nodes are moving around with an average velocity of 10m/sec. The transmission range for each node is, say, 350 meter. Under a given traffic pattern, will this network be able to provide the required service guarantee to its users in spite of the dynamic change in topology due to mobility? If the answer is yes, let us further assume that some of the users decide to switch-off or to leave the field or to increase their mobility. Will the network still survive? On the other extreme, let us assume that 20 new nodes join the system, making the node count to 40. The transmission range that is optimal for 20 nodes may be too high for 40 nodes, as this will increase collision and congestion of control / data packets. Will the network still be able to provide the required service guarantee to its users ?

So, survivability analysis and drawing up a specification for a survivable ad hoc network is an important issue that we want to address in this paper.

2. Survivable Systems

2.1 Definition and Characteristics of Survivable Systems

Traditionally, survivability in network systems has been defined as the capacity of a system to fulfil its mission, in a timely manner, in the presence of failures [7]. The term mission refers to a set of very high-level requirements or goals. Timeliness is a critical factor that is typically associated with the mission fulfillment. In the context of ad hoc network, mission fulfillment in a timely manner implies that the network should be able to ensure certain level of service guarantee to its user in the presence of system dynamics. Survivability analysis, in this context, can be defined as network specifications and management procedures to minimize the impact of system dynamics on the network services.

Thus, in this study, we are not considering failure due to hardware malfunctions or software errors. However, in an ad hoc network, any node can randomly switches itself off causing an event equivalent to node failure. Similarly, any link between two node can get disconnected anytime because of mobility of the nodes, causing an event equivalent to link failure. Additionally, new nodes can join the system at any point of time; similarly, new links can be formed between any two nodes, as they come closer to each other due to their mobility.

For survivability, we must achieve system-wide properties that typically do not exist in individual nodes. A survivable system must ensure that desired survivability properties emerge from interactions among the components in the construction of reliable systems from unreliable components [7]. If survivability properties are emergent they are present only when the number of nodes of a system are sufficiently large. If the number or arrangements of nodes falls below a critical threshold, the attendant survivability property fails. For example, we can specify an ad hoc network to operate at a transmission range of, say, 350 with number of nodes between 15 and 20 at a mobility ranging from 5m/s to 20m/s. But, if number of nodes falls below that, the system may not survive i.e. may not be able to ensure certain service guarantee to its users.

2.2 Specifying the Requirements of Survivable Ad Hoc Network

Central to the notion of survivability analysis is to identify and ensure the maintenance of certain essential attributes and the operating levels of those attributes that must be associated with the specified level of service guarantee. In the context of ad hoc network , the goal is to maintain the network availability and allow the data packets to be delivered to the intended destination from a source in spite of the changes in network topology due to its dynamic behavior. Survivability analysis consists of determining whether service objectives can be maintained during all operational modes.

Thus, network service in the context of ad hoc network is primarily pivotal to two fundamental requirements:
1. establishing a connection between any two nodes in the network at any instant of time.
2. Assuring an uninterrupted connection until a finite volume of data transfer has been accomplished (of course with limited delay in data transfer).

Survivability issues depend entirely on how well these two demands are met with. A network would be called survivable if it meets both the above requirements satisfactorily. Now, in order to declare an ad hoc network survivable we need to first define the user requirements in more formal terms. In other words we require a set of metrics which would inherently take care of all the service demands and finally throw up a numerical values depicting the degree of survivability for a given set of design specifications. Our objective is to design such a set of metrics in terms of the basic network parameters: number of nodes (N), transmission range (R), mobility (M), average volume of data to be communicated from a source to its destination (V) and average number of communication events per minute (C).

3. System Description

The network is modeled as a graph G = (N,L) where N is a finite set of nodes and L is a finite set of unidirectional links. Each node n ∈ N is having a unique node identifier. Since in a wireless environment, transmission between two nodes does not necessarily work equally well in both direction [1], we assume unidirectional links. Thus, two nodes n and m are connected by two unidirectional links $l_{nm} \in L$ and $l_{mn} \in L$ such that n can send message to m via l_{nm} and m can send message to n via l_{mn}. However, in this study, we have assumed $l_{nm} = l_{mn}$ for simplicity.

In a wireless environment, each node n has a wireless transmitter range. We define the neighbors of n, $N_n \in N$, to be the set of nodes within the transmission range R of n. It is assumed that when node n transmit a packet, it is broadcast to all of its neighbors in the set N_n. However, in the wireless environment, the strength of connection of all the members of N_n with respect to n are not uniform. For example, a node $m \in N_n$ in the periphery of the transmission range of n is weakly connected to n compared to a node $p \in N_n$ which is more closer to n. Thus, the chance of m going out of the transmission range of n due to outward mobility of either m or n is more than that of p.

Each link l_{nm} is associated with a signal strength S_{nm} which is a measurable indicator of the strength of connection from n to m at any instant of time. Due to the mobility of the nodes, the signal strengths associated with the links changes with time. When the signal strength S_{nm} associated with l_{nm} goes below a certain threshold S_t, we assume that the link l_{nm} is disconnected.

Affinity a_{nm}, associated with a link l_{nm}, is a prediction about the span of life of the link l_{nm} in a particular context [5]. For simplicity, we assume a_{nm} to be equal to a_{mn} and the transmission range R for all the nodes are equal. To find out the affinity a_{nm}, node n sends a periodic beacon and node m samples the strength of signals received from node n periodically. Since the signal strength is roughly proportional to $1/R^2$, we can predict the current distance d at time t between n and m. If M is the average velocity of the nodes, the worst-case affinity a_{nm} at time t is (R-d)/M, assuming that at time t, the node m has started moving outwards with an average velocity M. For example, If the transmission range is 300 meters, the average velocity is 10m/sec and current distance between n and m is 100 meters, the life-span of connectivity between n and m (worst-case) is 20 seconds, assuming that the node m is moving away from n in a direction obtained by joining n and m..

Given any path p = (i, j, k, ..., l, m), the **stability of path p** [5] at a given instant of time will be determined by the lowest-affinity link (since that is the bottleneck for the path) and is defined as $min[a_{ij}, a_{jk}, ..., a_{lm}]$. In other words, stability of path p between source s and destination d, η^p_{sd}, is given by

$$\eta^p_{sd} = min_{\forall i,j} \, a^p_{ij}$$

However, the notion of stability of a path is dynamic and context-sensitive. As indicated earlier, stability of a path is the span of life of that path from a given instant of time. But stability has to be seen in the context of providing a service. A path between a source and destination would be stable if its span of life is sufficient to complete a required volume of data transfer from source to destination. Hence, a given path may be sufficiently stable to transfer a small volume of data between source and destination; but the same path may be unstable in a context where a large volume of data needs to be transferred.

4. Route Discovery and Data Communication Mechanism in Ad Hoc Network

The existing routing protocol can be classified either as proactive or as reactive [3]. In proactive protocols, the routing information within the network is always known beforehand through continuous route updates. The family of distance vector and link state protocols is examples of proactive scheme. Reactive protocols, on the other hand, invoke a route discovery procedure on demand only. The family of classical flooding algorithms belongs to this group. It has been pointed out that proactive protocols are not suitable for highly mobile ad hoc network, since they consume large portion of network capacity for continuously updating route information. On the other hand, on-demand search procedure in reactive protocols generate large volume of control traffic and the actual data transmission is delayed until the route is determined.

Whatever may be the routing scheme, frequent interruption in a selected route would degrade the performance in terms of quality of service. In [5], we have attempted to minimize route maintenance by selecting stable routes, rather than shortest route, which is illustrated below.

4.1 Path Finding Mechanism

A source initiates a route discovery request when it needs to send data to a destination. The source broadcast a route request packet to all neighboring nodes. Each route request packet contains source id, destination id, a request id, a route record to accumulate the sequence of hops through which the request is propagated during the route discovery, and a count max_hop which is decremented at each hop as it propagates. When max_hop=0, the search process terminates. The count max_hop thus limits the number of intermediate nodes (hop-count) in a path.

When any node receives a route request packet, it decrements max-hop by 1 and performs the following steps:
1. If the node is the destination node, a route reply packet is returned to the source along the selected route, as given in the route record which now contains the complete path information between source and destination.
2. Otherwise, if max_hop=0, discard the route request packet.
3. Otherwise, if this node id is already listed in the route record in the request, discard the route request packet (to avoid looping).
4. Otherwise, append the node id to the route record in the route request packet and re-broadcast the request.

When any node receives a route reply packet, it performs the following steps:
1. If the node is the source node, it records the path to destination.
2. If it is an intermediate node, it appends the value of affinity and propagates to the next node listed in the route record to reach the source node.

4.2 Sending the Data from Source to Destination

When a source initiates a route discovery request, it waits for the route reply until time-out. If it receives a path, it computes its stability η^P_{sd}. If V_{sd} is the volume of data to be send to destination and if B is the bandwidth for transmitting data, V_{sd} / B is the one-hop delay to transmit the data, ignoring all other delay factors. If H is the number of hops from source to destination, $H* V_{sd} / B$ will be the time taken to complete the data transfer. If η^P_{sd} is sufficient to carry this data, the path is selected. Otherwise, the source checks the next path, if available, for sufficient stability. In order to check the sufficiency, η^P_{sd} is multiplied with a correction factor f, to be decided dynamically, to take care of estimation error and other delay factors related to traffic characteristics.

The Algorithm:
Step I: p:= 0;
Step II: **wait** for a path **until** timeout;
Step III: **if** a path is available **then**
begin

> p:=p+1;
> find η^P_{sd} = min $_{\forall i,j} \eta^P_{ij}$; { find the stability of path k}
> **if** $(H* V_{sd} / B) < f * \eta^P_{sd}$ {if the path is suffiently stable }
>   **then** start sending V_{sd} into p_{th} path
> **else** reject the path and goto step II

end
Step IV: terminate.

5. The Simulation Environment

Existing simulators are not well-equipped to serve our purpose [9,10,11]. Hence, in order to model and study the survivability issues of the proposed framework in the context of ad hoc wireless networks, we have developed a simulator with the capability to model and study the following characteristics:
- Node mobility
- Link stability (*affinity*)
- Affinity- based path search
- Dynamic network topology depending on number of nodes, mobility and transmission range
- Realistic physical and data link layers in wireless environment
- Data communication with different data volume and different frequency of communication events per minute.

The proposed system is evaluated on a simulated environment under a variety of conditions. In the simulation, the environment is assumed to be a closed area of 1000 x 1000 sq. meter in which mobile nodes are distributed randomly. We ran simulations for networks with different number of mobile hosts, operating at different transmission ranges. The bandwidth for transmitting data is assumed to be 1000 packets / sec. The packet size is dependent on the actual bandwidth of the system.

In order to study the delay, throughput and other time-related parameters, every simulated action is associated with a simulated clock. The clock period (time-tick) is assumed to be one millisecond (simulated). For example, if the bandwidth is assumed to be 1000 packets per second and the volume of data to be transmitted from one node to its neighbor is 100 packets, it will be assumed that 100 time-ticks (100 millisecond) would be required to complete the task. The size of both control and data packets are same and one packet per time-tick will be transmitted from a source to its neighbors.

The speed of movement of individual node ranges from 5 m/sec. to 20 m/sec. Each node starts from a home location, selects a random location as its destination and moves with a uniform, predetermined velocity towards the destination. Once it reaches the destination, it waits there for a pre-specified amount of time, selects randomly another location and moves towards that. However, in the present study, we have assumed zero waiting time to analyze worst-case scenario.

6. Analyzing the Impact of Dynamic Topology on Survivability

6.1 Related Definitions

To conceive certain trends in network characteristics on the whole, some terms have been used that are defined as follows:

Average Connectivity Efficiency (E): Connectivity Efficiency has been defined as the ratio of total number of connected node-pairs (in single hop or in multiple hops) and the total number of available node pairs at any instant of time. This fraction captures the degree of connectivity among the nodes in any snapshot of the mobile environment. From the survivability point of view, this parameter is an indicator to the success rate of a source node, in attempting to establish a connection with a destination node. The efficiency values obtained over several snapshots (taken at intervals of one second from the simulator) of the dynamic environment have been finally averaged to yield the Average Connectivity Efficiency. A network where all the node-pairs are always connected in single or multiple hops have a Average Connectivity Efficiency of 100%. Thus,

$$\text{Average Connectivity Efficiency } (\%) = \frac{\Sigma^T_{i=1} (\text{no. of connected node pairs}) * 100}{T * \text{Number of node-pairs}}$$

Average Network Stability (S): From survivability perspectives, the span of time for which two nodes remain connected (given the number of nodes, transmission range and the mobility) need to be analyzed. A parameter, **affinity**, introduced in [5] and explained in section 3 has been used for average worst case analysis. As explained in section 3, the stability of the path (i.e. the span of time for which this path would exist) can be determined by the weakest link in the path.

Two nodes in the ad hoc environment may often be connected with several paths. For data communication between two nodes, the best path should always be chosen i.e. the path assuring greater stability. Thus,

Node to Node Stability = *max* (stability of all the paths between the two nodes).

The **Average Network Stability** has been defined as the average node to node stability over time.

$$\text{Average Network Stability} = \frac{\sum_{i=1}^{T} \sum_{\text{all node-pair}} (\textbf{Node to Node Stability})}{T * \textbf{number of node-pairs}}$$

Average Number of Neighbors (G): The study of percolation is an important aspect from the data communication point of view in a mobile computing environment [12]. For a random distribution of nodes in a bounded region, percolation is proportional to the number of neighbors, which in turn is a function of node density and signal strength. Average Number of neighbors has been defined as:

$$\text{Average Number of Neighbors} = \frac{\sum_{i=1}^{T} \sum_{\text{all node}} (\textbf{Number of neighbors of each node})}{T * \textbf{number of node}}$$

6.2 Variation of Average Connectivity Efficiency (E) with N, R, and M

It is quite obvious that if the signal strength increases, the probability of connectivity also increases. The variation of connectivity efficiency against signal strength has been shown in the plot in fig.1(a). However this signal strength cannot be allowed to increase indefinitely due to other overheads:

1. Cost (power consumption due to battery usage) increases as the signal strength is raised.
2. Congestion and collision are the inevitable outcome of higher signal strength during data communication, as will be illustrated in the next section.

A larger number of users in a closed area indicate a higher node density. Since E is a measure of connectivity and connectivity is heavily dependent on how close the nodes are with each other, the total number of nodes in a bounded area also contributes to the connectivity efficiency. Thus, the connectivity efficiency bears a composite relation with the number of nodes as well (Fig. 1(b)). From figure 1, it is quite evident that, to achieve a specific threshold of efficiency, there is a lower cut off of the signal strength for a given number of nodes.

E would not depend on the mobility of the nodes. If the node mobility is high, then the probability of nodes moving out of a node's transmission range increases as much as the probability of new nodes coming into the transmission range of the same node. As a result, average value of connectivity taken over a long time remains unaffected at different mobility.

Figure 2 depicts the variation of Average Connectivity Efficiency (E) against Average Number of Neighbors (G). Although G does not reflect the actual dependence of E on N and R, it can be instrumental in deciding the cut off for satisfactory connectivity in the network. Over G=6 the efficiency is always found to

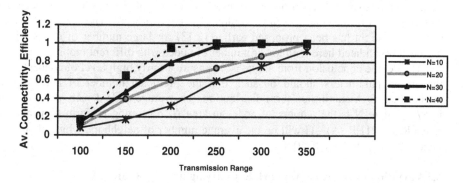

Fig 1(a). Average Connectivity Efficiency vs. Transmission Range for different number of nodes

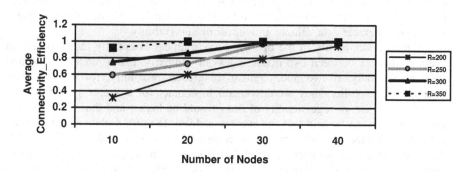

Fig 1(b). Average Connectivity Efficiency vs. Number of Nodes for different Transmission Range

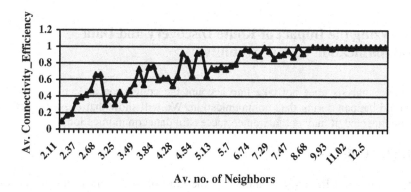

Fig 2. Average Connectivity Efficiency vs. Average Number of Neighbors

be increasing over 0.8. Over a certain threshold of neighbors, the network becomes connected and further increase in G would only hike overhead. The optimal value of G as six to eight has been proposed earlier[12,13] for large number of nodes which has been revalidated here with lower number of nodes with different mobility pattern.

Hence, the conclusion from above is: to ensure a fairly high level of connectivity, the design parameters should be such that the predicted number of neighbors is greater than 6. Assuming uniform distribution, the average node density per unit area is N/A, where N is the number of nodes in area A. Assuming uniform transmission range R, $[(N*\Pi R^2/A) -1]$ will be the average number of neighbors, which should be greater than six to have E >0.8.

6.3 Variation of Average Network Stability against N. R, and M

From the perspective of network service, the stability of a path between two arbitrary nodes indicates the volume of data that could be transferred between the two nodes in question (provided none of the intermediate nodes switch off during data transfer). Conversely, it is stability, which would be instrumental in deciding the thresholds of average transferable data volume, thus ensuring survivability.

A high node density in the operating environment essentially indicates that the average distance between two nodes is less in comparison to a model of low node density. Naturally, if two nodes remain in greater proximity, for a given signal strength and mobility, they would remain in contact for a longer period of time. Consequently the average stability of links would be higher and thus the stability of paths. Thus, it can be said that the average stability (S) of a mobile ad hoc wireless network would increase with increase in node density or N (as node density = N / A). At the same time, if the average affinity of links in a network features to be high, the average stability of paths would also be correspondingly higher. From the expression of affinity, we see that affinity of a link increases with increase in transmission range and/or decrease in mobility. Stability can thus be said to be directly proportional to transmission range and inversely proportional to mobility (Figure 3).

7. Analyzing the Impact of Route Discovery and Data Communication on Survivability

The above analysis does not take into account the congestion and collision factors that would happen during data communication. We will show that even if a network is well-connected, it may not guarantee successful data communication.

7.1 Related Definitions

Route Discovery Efficiency is defined as the ratio of the average number of route replies obtained per minute and the average number of route request generated per minute. As discussed, the number of route request generated per minute would depend on the number of communication events initiated per minute (C). However,

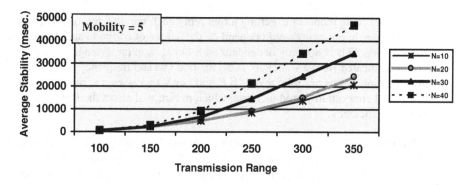

Fig 3(a). Average Network Stability vs. Transmission Range at mobility =5.

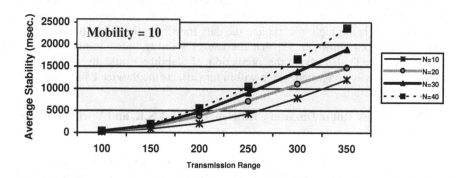

Fig 3(b). Average Network Stability vs. Transmission Range at Mobility = 10.

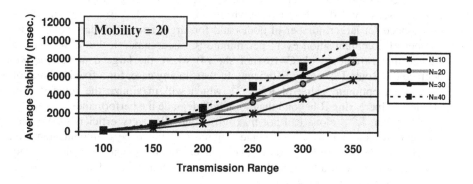

Fig 3(c). Average Network Stability vs. Transmission Range at Mobility =20.

the success of route request i.e. getting a rout reply back within a reasonable period of time (500 msec in our case) would depend on the degree of collision and congestion of the network. This is not only dependent on E but also on the average volume of data communicated from a source to its destination (V) and frequency of communication events per minute (C). If C and / or V increases, the probability of collision and congestion would increase, which in turn will affect the Roure Discovery Efficiency.

Service Efficiency is defined as the ratio of the average number of communication events successful within a reasonable period of time per minute and the average number of route request generated per minute. Service Efficiency depends on four factors : 1) Route request has been generated but route reply has not come back within a reasonable period of time, 2) Route replies have been obtained but the paths are rejected because they are not stable enough to carry out the required volume of data transfer, 3) A path is selected and data communication has started but the path could not be retained throughout the entire period of data communication, and 4) the network delay is too high to complete the data transfer within a reasonable period of time. It has been shown in [5] that the use of stability based routing reduces the probability of (3). However, the prediction of stability would be affected, if the network is heavily congested which will in turn affects the Service Efficiency.

7.2 Variation of Route Discovery Efficiency against N,R, and V with C=4 per Minute

For a given number of nodes, the number of control packets generated increases drastically beyond a certain transmission range. In a collision-free environment, if G is the average number of neighbors and max_hop =4, then the number of control packet generated will be G^4 per communication event. Therefore, it is obvious that with increase in G, the number of control packets increases drastically.

The congestion due to control packets at high transmission range would affect the Route Discovery Efficiency as shown in Figure 4. The effect would be more pronounced for larger number of nodes and for larger amount of data volume. Here, number of communication event per minute (C) is assumed to be 4. We have also studied this variation with C=10 (not shown) where the large data volume would degrade the Route Discovery Efficiency further. In any case, for a fixed number of N, there is an optimum value of R, R^{Nopt}, which will maximize the route discovery efficiency. Increasing R beyond that point will degrade the performance.

However, R^{Nopt} alone can not maximize route discovery efficiency. We need to consider two more factors : average volume of data to be communicated from a source to its destination (V) and average number of communication events per minute (C). The system should be capable of absorbing the control and data packets before a new communication event starts.

Depending on R^{Nopt} and the average network stability at that R, we can specify V for an average mobility M. If we increase M or V beyond that, the Service Efficiency will suffer.

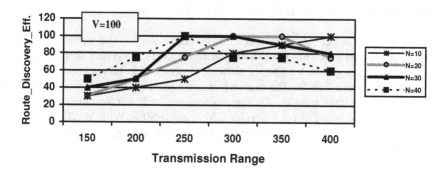

Fig 4(a). Route_Discovery_Efficiency vs. Transmission Range with
Data Volume = 100 packets, Max_Hop=4 and *No of Communication = 4 / min.*

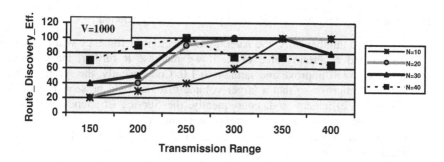

Fig 4(b). Route_Discovery_Efficiency vs. Transmission Range with
Data volume= 1000 packets, Max_Hop=4 and *No of Communication = 4 / min.*

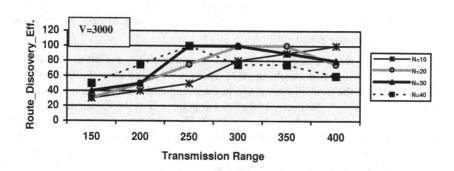

Fig 4(c). Route_Discovery_Efficiency vs. Transmission Range with
Data volume =3000 packets, Max_Hop=4 and No of Communication = 4 / min.

7.3 Variation of Service Efficiency with C=4 per Minute

For a given number of node and corresponding R^{Nopt} , the variation of Service Efficiency against M and V is shown in figure 5. It is evident that getting a high Service Efficiency with V=3000 is difficult to obtained in this set up where mobility is varying between 5 to 20. The reason is that we are not getting sufficient stable paths to complete the data transfer. On the other hand, for lower volume of data and low mobility, it is possible to get a Service Efficiency > 80%. With M=20, getting a high Service Efficiency for a data volume > 1000 is difficult, when the number of nodes are more than 20.

8. Survivability Metrics and Specifications for a Survivable Ad Hoc Network

From the above analyses, the following points can be concluded :

- For a fixed number of N, the Average Connectivity Efficiency will be more than 0.8 beyond a certain value of R. If we increase R further, the Connectivity Efficiency will improve and saturate to 1.0. Consequently, the Average Stability will also improve so that a larger volume of data could be sent. But the Route Discovery Efficiency and, consequently, the Service Efficiency will go down because of large number of control packets and / or data packets.
- For a fixed number of N, there is an optimum value of R, R^{Nopt}, which will maximize the route discovery efficiency.
- However, R^{Nopt} alone can not maximize route discovery efficiency. We need to consider two more factors : average volume of data to be communicated from a source to its destination (V) and average number of communication events per minute (C). The system should be capable of absorbing the control and data packets before a new communication event starts.
- Depending on R^{Nopt} and the average mobility M, we can specify average network stability which will in turn determine V. If we increase M or V beyond that, the Service Efficiency will suffer.

The aim of our entire analysis is to model the survivability region of operation for a mobile ad hoc wireless network. In other words, we need to answer questions like :
What should be the transmission range of operation and the maximum mobility for an ad hoc network with 30 users, if the user require a Service Efficiency of 80% and 1000-Kb average data volume for transfer? The kind of answers we are trying to provide is that, for 30 users with transmission range between 275 m to 325 m, it is possible to achieve the required Service Efficiency with V<=1000 and C=4, if the average mobility is less than 10. As another example, suppose we ask that: what is the Service Efficiency achievable if the number of users are 35 to 40, moving with an average velocity between 10 to 20 m/sec and the average data transfer requirement is 4 per minute with an average volume of data = 100 packets ? From the above analyses, we can say that with R=250, we can achieve a Service Efficiency of around 80%.

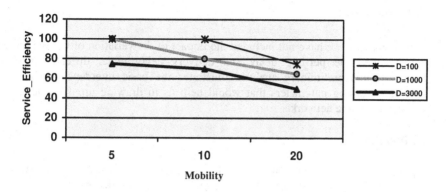

5(a). Service_Efficiency vs. Mobility with No. of Comm.=4/min. and _N=20_ & _R=350_ :

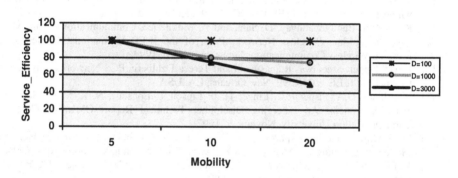

5(b). Service_Efficiency vs. Mobility with No. of Comm.=4/min and _N=30_ & _R=300_

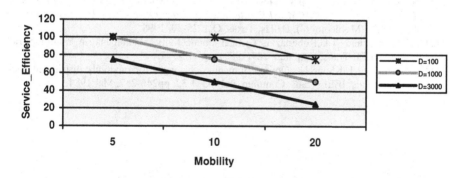

5(c). Service_Efficiency vs. Mobility with No. of Comm.=4 / min and _N=40_ & _R=250_.

9. Conclusion

In this analysis, we have not included the impact of the variation of C. We have also not included the per-hop delay and delivery delay under different conditions. However, this preliminary analysis illustrates the basic interdependencies among several governing parameters that would help us in drawing up specifications for survivable ad hoc networks.

Reference

[1] D. B. Johnson and D. Maltz, Dynamic source routing in ad hoc wireless networks, T. Imielinski and H. Korth, eds., *Mobile computing,* Kluwer Academic Publ. 1996.

[2] S. Corson, J. Macker and S. Batsell, Architectural considerations for mobile mesh networking, Internet Draft RFC Version 2, May 1996.

[3] Z.J.Haas, A new routing protocol for the reconfigurable wireless networks, ICUPC'97, San Diego, CA, Oct. 1997.

[4] V. D. Park and M. S. Corson, A highly adaptive distributed routing algorithm for mobile wireless networks, Proc. IEEE INFOCOM '97, Kobe, Japan, April 1997.

[5] K. Paul, S. Bandyopadhyay, D. Saha and A. Mukherjee, Communication-Aware Mobile Hosts in Ad-hoc Wireless Network, Proc. of the IEEE International Conference on Personal Wireless Communication, Jaipur, India, Feb. 1999.

[6] David Tipper, Sreeniwas Ramaswami and Teresa Dahlberg, PCS Networks Survivability, to appear in IEEE WCNC 99, New Orleans, LA, USA

[7] R.J.Ellison D.A. Fisher R.C. Linger H. F. Lipson T. Lonstaff, N. R. Mead, Survivable Network System: An Emerging Discipline, Technical Report CMU/SEI-97-TR-013, Carnegie Mellon University, November, 1997.

[8] D. Medhi, A Unified Approach to Network Survivability for Teletraffic Networks: Models, Algorithms and Analysis, IEEE Transactions on Communications April 1994.

[9] K. Fall, and K. Varadhan. ns Notes and Documentation. The VINT Project, UC Berkeley. http://www-mash.cs.berkeley.edu/ns/,1997.

[10] J. Short, R. Bagrodia and L. Kleinrock. „Mobile Wireless Network System Simulation." *Wireless Network Journal 1*, no. 4, 1995.

[11] Josh Broch,; D.A. Maltz; D.B. Johnson; Y. Hu and J. Jetcheva. „A Performance Comparison of Multi-Hop Wireless Ad Hoc Network Routing Protocols." In *Proceedings of the Fourth Annual ACM/IEEE International Conference on Mobile Computing and Networking* (Mobicom'98), Dallas, Texas, Oct. 25-30, 1998.

[12] Y.C.Cheng and T.G.Robertazzi, Critical connectivity phenomena in multihop radio network, IEEE Trans. Commun. , 37(1989), pp 770-777.

[13] H.Takagi and L.Kleinrock, Optimal transmission ranges for randomly distributed packet radio terminals, IEEE Trans. Commun. vol COM-32, pp246-257, Mar.1984.

A Token Passing Tree MAC Scheme for Wireless Ad Hoc Networks to Support Real-Time Traffic

Rao Jianqiang[1], Jiang Shengming[2], and He Dajiang[1]

[1] Department of Electrical Engineering, National University of Singapore,
10 Kent Ridge Crescents, Singapore, 119260
{engp9040, engp8587}@nus.edu.sg
[2] Centre for Wireless Communications, National University of Singapore,
20 Science Park Road, #02-34/37, Singapore, 117674
jiangsm@cwc.nus.edu.sg

Abstract. This paper introduces a distributed MAC protocol which can provide delay-bounded service in wireless ad hoc networks. Ad hoc networks are network architectures that do not rely on a pre-existing fixed infrastructure, which imposes heavy challenges on designing MAC scheme with QoS support. Most existing MAC protocols for wireless ad hoc networks have been designed to support none-delay-sensitive applications. Given an increasing demand on supporting multimedia applications, there is much interest in designing MAC schemes to satisfy the requirements of real-time applications in ad hoc environments. The timed token based MAC scheme proposed in this paper allocates the bandwidth to the users according to their requirements. This scheme can guarantee the deadline of real-time traffic even in the case of heavy traffic load. In addition, a logical token-passing tree structure adopted in this scheme overcomes the hidden terminal problems in ad hoc networks.

1   Introduction

Since there is no need of preexisting infrastructure, a wireless ad hoc network can be rapidly deployed. Thus, such networks which can operate in different network conditions provide a low cost and flexible solution to communication networks. Most interest in this technology comes from both the commercial market and military applications. There are many challenges of the current research in this area. One of those issues is medium access control (MAC) which controls nodes to access medium and determines quality of service (QoS) and channel utilization [1]. Most existing MAC protocols which can provide QoS capability to users require some central entities, e.g., the base-stations in cellular wireless networks. Such schemes are not suitable in ad hoc environments because of the lack of central entities as mentioned above. Although a node can be selected to function as a central unit temporarily, this node cannot perform as robustly as a fixed one because of the possibility of platform movements. Therefore, the distributed MAC schemes are more desirable than the centralized MAC schemes used in cellular systems, although the latter support real-time traffic efficiently under the control of the central entity.

C.G. Omidyar (Ed.): MWCN 2000, LNCS 1818, pp. 47–57, 2000.

Carrier sense multiple access (CSMA) is one of the most pervasive distributed MAC scheme used in wired network. But, Packet collisions are intrinsic to CSMA due to the nature of carrier sensing and the existence of hidden terminals in ad hoc networks. Many MAC schemes [2,3,4] either with or without carrier sensing, including the IEEE802.11 standard [5], have been proposed to improve the throughput over that of CSMA. These schemes handle collision avoidance by exchanging an in-band control packet (RTS/CTS) handshake prior to data transmission. The sender can transmit a data packet only after a successful RTS/CTS exchange. And collisions are solved by backing off and rescheduling RTS transmissions. These protocols, however, cannot meet the delay requirements of real-time traffic. The real-time packets may not be transmitted before their deadline because the control packets still suffer from collisions and their retransmissions are randomly scheduled.

Recently, several novel distributed MAC schemes have been proposed, aimed at providing the QoS guarantees for real-time traffic support. With the scheme proposed in [6], real-time nodes contend for access to channel with pulses of energy, the durations of which are function of the delay incurred by the nodes until the channel becomes idle. The resulting scheme guarantees priority to real-time traffic and provides round-robin service and bounded access delay to real-time nodes. Another scheme [7] uses window splitting protocols with limited feedback sensing to offer support for deadlines of real-time data. However, the two schemes mentioned above do not consider the existence of hidden terminals. The two other reservation-based schemes [8,9] can guarantee the contention-free transmission based on the exchange of channel state informations among the active nodes. Of course, as the wireless spectrum is premium, the frequent exchange of network informations which may cause much overhead should be avoided.

In this paper, we introduce a Token Passing Tree (TPT) structure for indoor ad hoc networks, in which terminals have low mobility because of low movement speed (i.e., walking speed) and limited movement space. The possible scenarios include meeting rooms, wireless offices and other indoor environments where wiring is to be kept to minimum. A temporary ad hoc network can be rapidly constructed and used by users to exchange informations in such situation. The proposed MAC scheme is based on timed token protocol [10]. The timed token protocol is suitable for real-time communications not only because of its contention free nature but also due to the fact that it has the important property of bounded access time which is necessary for real-time communications. It is used pervasively in the wired network. However, the challenges on using the token-passing structure in the wireless ad hoc network comes from three aspects: the unreliable characteristics of wireless medium, the hidden terminals and the dynamics of nodes. The dynamics of nodes means that each node may switch off or on unpredictably. Addressing this, a Token Passing Tree structure is adopted, which is more easily maintained than traditional token passing scheme. Furthermore, a mechanism is provided to reconstruct the TPT when the token passing process is hampered due to the lost of the token. As long as a node maintains its position in a TPT, the QoS requirements of this node can be guaranteed.

In Section 2 we describe the MAC scheme and derive the condition to guarantee the bounded access delay to real-time node. The maintenance of TPT is also addressed in this section. Then, Section 3 presents the performance evaluation of proposed scheme. Finally, we conclude the paper in the last section.

2 MAC Description

The wireless ad hoc networks under consideration consist of N nodes, approximately less than 30, all of which are functionally identical. Each node is trying to communicate with one another over a single hop, or with a access point or cluster head to reach a hidden nodes. In this case, each sender is within sight of the access point or cluster head. This is reasonable assumption for the hierarchical ad hoc networks, in which the network nodes are partitioned into clusters. As depicted in Figure 1, a typical TPT structure looks like a bi-directional tree with the token-rotation beginner (TRB) being the root. There are two logical token-passing directions: forward and backward. The former is used by a parent-node forwarding a token to its child-nodes while the latter is used by a child-node returning the token to its parent-node. The maintenance of TPT is presented in Section 2.2. Some parameters used in the following discussions are defined below.

- T_{prop} is the propagation delay between a pair of nodes which is supposed to be identical to all nodes.
- T_{proc} is the transmission time of a token frame.
- H_e is the total reserved bandwidth of all nodes in TPT. $H_{e,i}$ is the bandwidth reserved by node i.
- D is the common bounded access delay of nodes in TPT. D_i is the bounded access delay of node i.
- N is the total number of nodes in TPT.

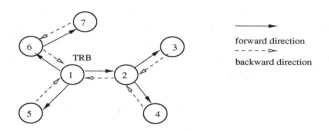

Fig. 1. Typical structure of TPT

A token frame contains the following information: a token sender ID (SID), which represents the identity of the node transmitting the packet; a token receiver ID (RID), which represents the identity of the node invited to transmit next; a Random Access Segment (RAS) flag, which is used in the procedure of node joining TPT; the configuration parameters of TPT (D,N,H_e), which is used in TPT configuration. The token frame can be carried in the data packets to decrease the overhead caused by the token transmission. When a current token-holder transmits it last data packets, it add the ID of the node invited to transmit next in this packet. If the token-holder has no packet to send, a individual token frame is transmitted to the intended receiver in TPT. The ID of each node is given when it joins the TPT. After a node receives the token from its parent-node, it first detects whether the RAS flag is set. If the RAS flag is set, this node must wait a random access period T_{rap} which is used by new nodes to submit request to join TPT. Then, it begins to transmit its data packet. The RAS flag is dynamically set to achieve the high channel utilization. In this scheme, we define that the RAS flag is set once every two cycles. And the nodes in TPT take their turns to set the RAS flag in the token frame.

2.1 Bounded Access Delay with TPT

Based on the timed token MAC protocol, a formed TPT can provide bounded access delay which is necessary to ensure that the transmission deadline of the real-time traffic is satisfied. The TPT adopts the mechanism used by the timed token protocol to allocate available bandwidth to the nodes. The timing properties of the timed token protocol shows that the average token rotation time is bounded by the Target Token Rotation Time (TTRT) and maximum token rotation time cannot exceed twice the TTRT. During the network configuration process, all nodes negotiate a common value for the bounded access delay D, an important parameter which guarantees the transmission deadline of the real-time traffic. Since each node has different bounded access delay requirements (denoted as D_i for node i) to be satisfied, the negotiated bounded access delay D should be the minimum value of all the D_i. that is,

$$D = min\{D_i, \forall i \in \Psi_{TPT}\} \tag{1}$$

where Ψ_{TPL} is the set of the nodes in a TPT.

Each node can reserve an amount of time (denoted as $H_{e,i}$ for node i) for transmission every time it receives a forward token. To guarantee bounded access delay D to all nodes, based on the timed token protocol, the sum of $H_{e,i}$ allocated to each node can not exceed the Target Token Rotation Time, which equals the half of D. Therefore, the bandwidth allocation scheme can be expressed as follows,

$$\sum_{\forall i \in \Psi_{TPL}} H_{e,i} + 2(N-1)(T_{proc} + T_{prop}) + T_{rap} \leq \frac{D}{2} \tag{2}$$

Once the node i receives the forward token, the minimal time it can use is $H_{e,i}$. Then it can uses the left bandwidth only if the time elapsed since the previous token arrival is less than the value $\frac{D}{2}$. The amount of time (denoted as H_i) token holder i can use is,

$$H_{max,i} = \begin{cases} H_{e,i} + \frac{D}{2} - (T_{a,i} - T_{la,i}) \text{ if } (T_{a,i} - T_{la,i}) \leq \frac{D}{2} \\ H_{e,i} \hspace{3cm} \text{else} \end{cases} \quad (3)$$

where $T_{a,i}$ is the forward token arrival time of node i, $T_{la,i}$ is the forward token last arrival time of the same node.

2.2 TPT Configuration

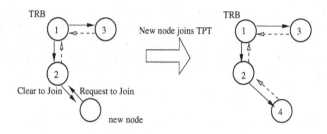

Fig. 2. Node joining procedure

Call Admission Control When a node becomes active, it first listens to the channel. If there is no TPT undergoing, this node can act as a TRB and construct a TPT with one node. Otherwise, it joins the current TPT as a new node. The new node can request to join the TPT only during the random access period (RAP). The duration of random access period T_{rap} is longer than the sum of the length of a request to join (RTJ) and a clear to join (CTJ), plus a maximal round-trip propagation time. Figure 2 shows the procedure of a new node joining TPT. Suppose now, the node 2 in TPT returns the token to its parent-node 1. The node 1 receives the token and waits T_{rap} for new node joining TPT. Almost simultaneously, the new node can capture this token because the transmission of token is over radio channel, and get the address of node 2 from the token. Thereafter, it sends a RTJ to the node 2 during the RAP. The RTJ carries the bandwidth requirement of the new node ($D_{new}, H_{e,new}$). The node 2 (landing node) receives the RTJ from the new node and performs the call admission control. This control determines whether the new call is admitted according to the requirement of new node, subject to the bounded access delay guaranteed by the current TPT. A new admission can occur if and only if the bounded access delay of the current TPT can be maintained. i.e,

$$\overbrace{H_e + H_{e,new}}^{H_{new}} + 2(\overbrace{N+1}^{N_{new}}-1)(T_{proc} + T_{prop}) \leq \frac{\overbrace{min\{D, D_{new}\}}^{D_{new}}}{2}. \quad (4)$$

Given that the condition is satisfied, the node 2 sends a clear to join (CTJ) to the new node and a unique ID is endowed with the new node. Then, the node 2 invokes the TPT configuration process (TPTCP) to inform the remaining node in TPT when it receives the token from the node 1 the next time. The token is set to be a configuration token which carries the new D_{new}, N_{new} and H_e. The RAS flag in the token is disabled, which prevents other new nodes from sending the RTJ during the TPTCP. The TPTCP goes along with the normal token passing path. All other nodes in TPT can get the D_{new}, N_{new} and H_e from the configuration token. After the TPTCP token finishes its rotation, the node 2 forwards the token to the new node and the new nodes is added in the new TPT. The procedure of node joining TPT is parallel with the normal token passing operation which has no impact on the performance of this scheme except the overhead incurred by the RAP. It is possible that two or more new nodes sent the request to join the TPT simultaneously. Thus, the RTJs sent by these nodes will collide. To resolve the collisions, the RTJs are randomly scheduled to be retransmitted if the new nodes do not receive the CTJs from the landing node in TPT.

2.3   Token Loss

Communication failures in the wireless environment are a relatively frequent occurrence. With the proposed MAC scheme, the communication failures may be caused by the token passing failures, due to transmission error and/or node's powering off. For the token passing structure to be feasible in such environments, the TPL should be recovered without causing much overhead. In the case of a fixed network coordinator, the polling scheme is adopted. While in a distributed system each node needs to achieve a common decision on who should be responsible to recover the TPL in the case of token passing failure. For this reason, this scheme adopts a timer to detect the token passing failures and ensure that the TPL is recovered. This timer is set by each node in a distributed manner. The allocation scheme given in Section 2.1 can guarantee the interval between two successive token's arrivals at a node shorter than the common bounded access D. Thus a node sets a timer T_{timer} equal to D each time it receives the token. Then, T_{timer} counts down until the token arrives at this node next time. If the token arrives at this node before its timer expires, T_{timer} is reinitialized to D and enabled again(starting the counting down process). If the token does not arrive before its timer expires (i.e., $T_{timer}=0$), it means that the token passing failure happens. This node regenerates a token and continues the token token passing loop. If it receives the token in the following loop, the TPT is recovered. Otherwise, the link between this node and its parent node is considered being broken. Then, it becomes a TRB and broadcasts this message to the other nodes. The broadcasted message can reach all hidden nodes of this node with the help of cluster or the access point. Then, the original TPT scheme is switched to the contention based MAC scheme, each node can join the TPT as a new node using the exchange of RTJ/CTJ with the nodes in new TPT. Because each node has its unique timer, only the node which first detects the token lost is chosen as TRB.

Considering the case of a network without the cluster, the broadcasted message may not reach all nodes in TPT before their timer expire. Thus, it is possible that two or more nodes become the TRB because their timer expire. Then, several TPTs may be constructed by these TRBs. We call this issue multi-TPT. The multi-TPT also exits in the multi-hop ad hoc networks. With the multi-TPT, the better space reuse can be achieved. But, at the same, it exits the problem that the token rotation of one TPT may interfere with the token rotation of its neighbor's TPT. Therefor, the maintenance of the multi-TPT is a much more complex open problem and is my undergoing work.

3 Performance Evaluation

The channel capacity of channel in wireless ad-hoe network is less than 10Mb/s, while the timed token protocol is generally adopted in the high bandwidth wired network(100Mb/s). Thus, we have simulated this scheme over a variety of network configurations. The overall performance of this scheme in the steady state with a mixed population of data and real-time nodes is studied with simulations. We mainly investigate the effect of the two parameters, the bounded access delay of real time traffic and the data packet size. There are 10 real-time nodes, 20 data nodes in our simulation model. Table 1 shows the traffic model of real-time and data nodes. The channel bit rate r_c of system is 2Mb/s. The traffic arrival rate r_s of each real-time node is 64kb/s. We assume that real-time packets are presented to the MAC layer periodically. Thus, the assembled packet size (b_{rpkt}) is the product of the inter arrival rate (A_r) of the real-time packet and the source rate r_s. The bounded access delay D equals the deadline of real-time packet, which equals A_r. The reserved bandwidth of each real-time node equals the transmission time of a real-time packet. We can calculate that the total reserved bandwidth of real-time nodes is less than $\frac{D}{2}$. The other parameters are shown as follows. The size of token frame is 20 bytes. The T_{rap} is 0.1ms. The propagation delay T_{prop} is 0.1us. The data load is defined as $N_{data}\lambda b_{pkt}/r_c$, where N_{data} is the number of data station. Because the packet size is fixed, the delay is defined as the access delay. The throughput of data traffic is measured as the normalized data packets transmitted over the channel.

Because the real-time nodes are guaranteed with bounded access delay D in the steady state as shown in previous section, it shows that the maximal real-time packet delay is less than the deadline of packet. No real-time packets are dropped even in high data load. Figure 3 shows the average real-time packet delay against the data load. The assembled real-time packet size is 1600bit, 3200bit and 6400bit respectively. Accordingly, the bounded access delay D is 25ms, 50ms and 100ms. The result shows that the average real-time packet delay is less than one third of the deadline of the packet even in the high data load. Given the same data load, the delay of real-time packet increases as its size increases. Figure 4 shows the throughput of data traffic against the data load. The contention free

Parameters	Real-time node	Data node
Interarrvial pdf	Constant	Poisson
Interarrvial rates(ms)	A_r	λ
Packet length pdf	Constant	Constant
Mean packet size(bytes)	b_{rpkt}	b_{pkt}
Reserved bandwidth	b_{rpkt}/r_c	0
Packet deadline(ms)	D	no requirement

Table 1. Traffic models

nature of token passing scheme ensures that every transmission is successful in a distortion free channel. The throughput can therefor reach a maximal value and is not degraded under heavy loads. We can see from Figure 3 and 4 that there is a tradeoff between the QoS requirements of real-time traffic and the throughput of the data traffic. The throughput of data traffic decreases as the bounded access delay D decreases when the network is over loaded, while the smaller value of the bounded access delay means better QoS that the real-time traffic can get. The reason is that the transmission of data packet is constrained by the bounded access delay [11]. Similarly, Figure 5 shows the impact of guaranteeing priority to real-time traffic on the average delay of data packet. The average data packet delay increases more quickly with the larger D as the data load increases.

Figure 6 and 7 compares the average data packet delay and throughput of data traffic between the cases of different fixed-length data packet. We observe that the performance of this scheme is better when the data packet are shorter. The shorter data packet is more desirable to achieve better performance.

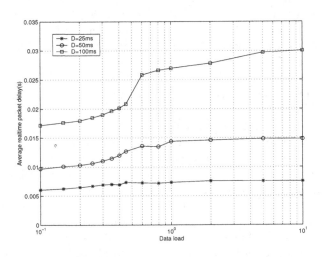

Fig. 3. Real-time packet delay versus data load

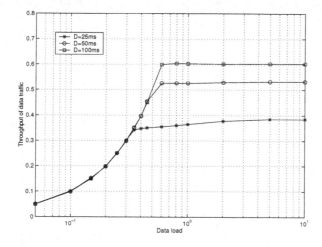

Fig. 4. Throughput of data traffic versus data load

4   Conclusions

We have described the structure of the token passing based MAC scheme for ad hoc networks, and investigated its major performance through simulation. The result have shown that this scheme can effectively support real-time traffic in the steady state. The ongoing work is to investigate the impact of the token lost events on this scheme and do the detailed performance analysis of this model. Our future studies will focus on improving the adaptivity of current MAC scheme.

References

1. M. Conti, C. Demaria and L. Donatiello.: Design and Performance Evaluation of A MAC Protocol for Wireless Local Area Networks. *ACM Mobile Networks and Application,* Vol. 2, No. 1, 1997, pp69-87.
2. P.Karn, MACA-A new channel access method for packet radio.: *Proc. ARRL/CRRL Amateur Radio Ninth Computer Networking Conf,* 1990.
3. F. Talucci and M. Gerla.: MACA-BI (MACA By Invitation) A Wireless MAC Protocol for High Speed Ad hoc Networking. *The 6th IEEE International Conference on Universal Personal Communications,* 1997, pp913-917.
4. C.L. Fullmer and J.J. Garcia-Luna-Aceves.: Floor Acquisition Multiple Access (FRMA) for Packet-Radio Networks. *Proceedings of ACM SIGCOMM,* 1994.
5. IEEE 802.11. Draft Standard for Wileless LAN. P802.11D5.0, 1996
6. J.L.Sobrinho and A.S.Krishnakumar.: Real-Time Traffic over the IEEE 802.11 Medium Access Control Layer. *Bell Labs Technical Journal,* Vol: 1, No.2, Autumn 1996, pp172-187.

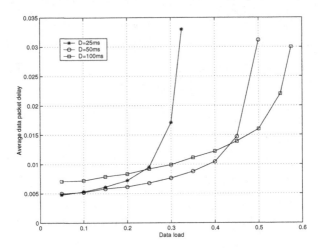

Fig. 5. Data packet delay versus data load

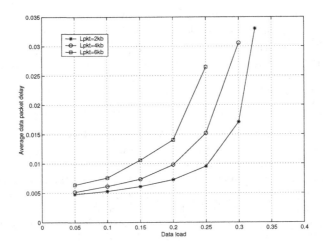

Fig. 6. Data packet delay versus packet length

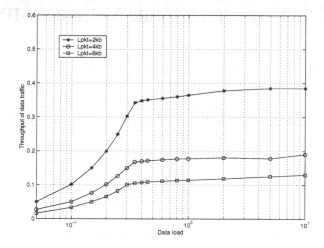

Fig. 7. Throughput of data traffic versus the data load

7. Michael J.Markowsk and Adarshpal S.Sethi.: Fully Distributed Wireless Transmission of Heterogeneous Real-Time Data. *Vehicular Technology Conference,* Volume: 2, 1998, pp1439-1442.
8. C.R. Lin and M. Gerla.: Asynchronous Multimedia Multimedia Radio Network. *IEEE INFOCOM'97,* 7-11 March, 1997.
9. A.Muir and J.Garcia-Luna-Aceves.: Supporting real-time multimedia traffic in a wireless LAN. *Proc. SPIE multimedia Computing and Networking,* pp.41-45 1997.
10. R.M.Grow.: A timed token protocol for local area networks. *Proc.Electro/82,Token Access Protocols,* May 1982.
11. Seung Ho Hong.: Approximate analysis of timer-controlled priority sceme in the single-service token-passing system. *Networking,IEEE/ACM Transaction,* Volume:22, April 1994, pp.206-215.

Challenges for Mobile Voice-over-IP

Prathima Agrawal[1], Jyh-Cheng Chen[1], and Cormac J. Sreenan[2]

[1] Applied Research, Telcordia Technologies, Morristown, NJ 07960, USA
{pagrawal,jcchen}@research.telcordia.com
[2] Department of Computer Science, University College Cork, Cork, Ireland
cjs@cs.ucc.ie

Abstract. The use of IP to transport voice is receiving significant attention in the telecommunications and data networking industries. Commonly known as Voice-over-IP (VoIP), this technology has the potential to allow user communication involving multiple media types, and between terminals that offer improved capabilities and user interfaces. Meanwhile there has been explosive growth in the demand for wireless connectivity, raising the question of how VoIP can be extended to situations where users and terminals can be mobile. In this paper we survey the technologies necessary to provide mobile VoIP, and identify relevant technical challenges.

1 Introduction

IP telephony is expected to see continued growth in supporting corporate and commercial services. This growth points to a large business and market value for Voice-over-IP (VoIP) in the near future. Currently IP telephony provides voice service to end terminals that are attached to wired networks. With the proliferation of mobile and wireless services and as more users disconnect from their fixed access points and become mobile, there is an increasing need to adapt VoIP to the mobile and wireless domain [10].

However, the mobile environment is much more dynamic and subject to change than the traditional wireline environment. The uncertainties of the wireless and mobile environments call for an increased level of adaptability. For more robust real-time communications, a signaling protocol is vital in providing highly reliable and robust connectivity in such communications environment. In addition to establishing and releasing a call, a signaling protocol may also need to monitor and maintain connectivity when the end terminal is moving and/or the transmission capabilities are varying. Dealing with issues of user and terminal mobility in an internetwork takes on special significance when voice transport is involved, because of latency and general performance constraints that are not as important for non-real time communication. Similarly, the service quality necessary for voice communication demands packet scheduling and resource reservation techniques which can operate effectively in the presence of mobility.

This paper is organized as follows. Section 2 provides a survey of signaling for VoIP, and describes a signaling architecture for use with mobile VoIP terminals.

C.G. Omidyar (Ed.): MWCN 2000, LNCS 1818, pp. 58–69, 2000.

Section 3 explains how user and terminal mobility are handled in the telephone system and in the Internet. Performance issues for supporting voice to mobile IP terminals are discussed. Section 4 discusses how to provide support for quality of service (QoS) for packet voice communication over wireless links and where terminals are non-stationary. Finally, Section 5 concludes the paper.

2 Signaling

The first challenge of IP telephony is to initiate interactive communication sessions between users. Two major standards have recently emerged for the signaling and control of VoIP. The first is H.323 [8] supported by the ITU-T, and the second is the Session Initiation Protocol (SIP) [6] proposed by the IETF. These two protocols provide similar signaling functionality, however H.323 is widely deployed in commercial products. Microsoft released for example NetMeeting, a H.323-compliant product, in 1996. On the other hand, SIP is a lightweight protocol which provides a simpler implementation and a greater degree of efficiency than H.323. These two protocols, however, currently do not provide support for resource reservation and QoS guarantees. DOSA (Distributed Open Signaling Architecture) [5] and TOPS (Telephony Over Packet networkS) [9], proposed by AT&T, provide a framework to integrate call signaling and resource management for IP telephony, and to enable creative new services for packet telephony. ITSUMO [7], a Telcordia-Toshiba joint research project, aims on providing IP signaling along with terminal mobility and QoS support.

2.1 H.323

Originally H.323 was developed for visual telephone terminal conferencing over non guaranteed QoS LANs and is an umbrella standard covering audio and video coders, call signaling, connection control, data and conference control, media transport, etc. In H.323, the signaling functionality is migrated to end terminals which are intelligent end points instead of the dumb end points used in the PSTN (Public Switched Telephone Network). H.323 is also not tied to a single transport mechanism, and can run over a variety of networks, including ATM and ISDN.

Fig. 1 shows a typical architecture and components of an H.323 LAN which is called a H.323 zone. The terminal in H.323 usually is a multimedia PC. The gateway (GW) is an endpoint on the network which provides for real-time, two-way communications between H.323 terminals on the packet-based network, other ITU terminals on a switched circuit network, and other H.323 gateways. The Multipoint Control Unit (MCU) is an endpoint which provides the capability for three or more terminals and gateways to participate in a multipoint conference. The gatekeeper (GK) is an entity that provides address translation and controls access to the network for H.323 terminals, gateways, and MCUs. The gatekeeper may also provide other services to the terminals, gateways and MCUs such as bandwidth management and locating other gateways.

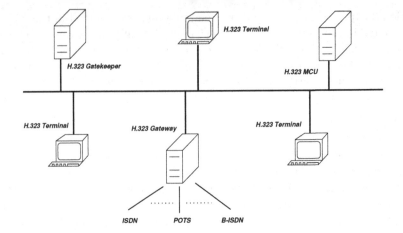

Fig. 1. H.323 zone

H.323 uses H.225.0 as the connection establishment protocol and H.245 as the control protocol between H.323 clients to establish a call, negotiate terminal capability and open logical channels. In the call that involves gatekeepers, a terminal that wants to set up a call first sends a request to its GK. After the GK exchanges message with the other GK, the GK responds to the call originator (caller) with the address of the desired destination (callee). The caller then sends the request directly to the callee. The callee then asks its GK for permission after sending a call proceeding message to the caller. Once the permission is granted by its GK, the callee responds to the caller. After the negotiation between the end points, the media channels are opened and the terminals then exchange media data using RTP/RTCP.

To provide terminal mobility for H.323-based IP telephony, [11] proposes leveraging the dynamic join and departure of multipoint conference with location update. It supports mobility for H.323 terminals without adding new entities, and with minimal modifications to the standard.

2.2 SIP

SIP (Session Initiation Protocol) is a signaling protocol for IP telephony developed by the IETF. SIP originated out of a lightweight Internet approach and reuses many of the header fields, encoding rules, error codes and authentication mechanisms of HTTP. As with HTTP, SIP operates independently of the packet layer and uses text to encode its messages, which facilitates parsing and helps to simplify debugging.

To place a call in SIP, the call originator (caller) locates an appropriate SIP server, typically by DNS. The caller then sends an INVITE request to the SIP server. The handling of INVITE depends on the nature of the server. In the case of a proxy server, the server queries a location server to find the address of the desired destination (callee), and forwards the INVITE message to the callee. The callee sends an ACK to the proxy server which forwards it back to the caller. All subsequent signaling between caller and callee is then handled through the proxy server.

In the case of a redirect server, the server queries the location server to find the callee, then supplies the location of the callee to the caller. The caller then issues an INVITE message to the callee. All subsequent signaling is direct. SIP can also be used simply from a client to another client. SIP deals with user mobility by relying on location servers, but was not designed to address the issues of terminal mobility. Section 3 discusses a solution that uses SIP to provide terminal mobility.

2.3 DOSA and TOPS

Traditional circuit telephony service ensures that end-to-end resources are reserved before the called party's phone is made to ring. However, both H.323 and SIP are signaling protocols without any consideration of resource reservation and QoS guarantees. DOSA (Distributed Open Signaling Architecture), which is a framework for call signaling and resource management for IP telephony proposed by AT&T, provides similar service to conventional telephony. In DOSA, telephony clients, such as PCs or multimedia terminal adaptors, participate in end-to-end capability negotiation, call signaling, and resource reservation. DOSA incorporates an explicit coordination between the call signaling protocol and the resource management protocol such that users are authenticated and authorized before receiving access to the enhanced QoS associated with the telephony service.

TOPS (Telephony Over Packet networkS) also proposed by AT&T, on the other hand, allows users to use mobile terminals or to move between terminals but still can be reached by the same user distinguishing name. Users can control how calls are routed to them. Terminals in TOPS have a range of capabilities to support different media. The major elements of TOPS include a directory service, an application layer signaling (ALS) protocol, a logical channel (LC) abstraction, and mechanisms to support a variety of conferencing modes. The directory service provides flexible mapping of a user's name to a set of terminals where the user can be reached. The ALS negotiates capabilities, establishes and maintains call state, and support advanced features. The LC abstraction isolates the applications from the underlying networks. The TOPS system has been implemented and is being used experimentally within AT&T Labs - Research.

2.4 ITSUMO Signaling

ITSUMO (Internet Technologies Supporting Universal Mobile Operations), a Telcordia and Toshiba joint research project, envisions an end-to-end wireless/ wireline IP platform for supporting future real-time and non-real-time multimedia services. The goal is to use IP and third (and above) generation wireless packet technologies to design a wireless platform that allows mobile users to

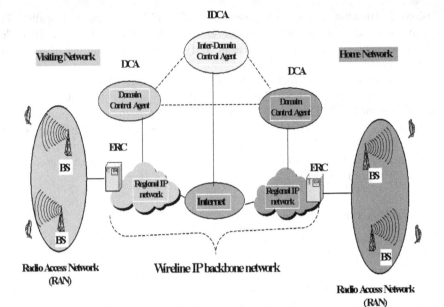

Fig. 2. ITSUMO's long term network architecture

access all services on a next generation Internet. Fig. 2 depicts the end-to-end packet platform of ITSUMO's all IP wireless/wireline network, which comprises all IP wireless access networks and an IP backbone network. The backbone network is an end-to-end wireline IP infrastructure that will comprise regional providers' wireline networks that are connected through the Internet. Besides mobile stations/terminals, a wireless access network also comprises a radio access network (RAN), and an edge router and controller (ERC). In order to support the needs of its users, a wireless access network interacts with the network control entities that are shown as Domain Control Agents (DCA) in Fig. 2.

The long-term control architecture of ITSUMO is shown in Fig. 3. It uses SIP for its end-to-end control (e.g., DCA, IDCA, etc.) architecture. All mobile stations (MSs) and fixed hosts have SIP user agents that interact with the SIP servers (i.e., proxy servers, redirect servers, and registrar) of the DCA/IDCA within the network. In Fig. 3, the SIP server entity of a regional IP network DCA represents the set of SIP proxy and SIP redirect servers within the regional network that perform the network control and signaling functions. Similarly, the registrar represents the server and database (or set of servers and databases) that accepts (accept) SIP REGISTER requests and provides (provide) location services that are similar to those of the home/visiting location registries (HLR/VLR) in today's cellular telephony. As Fig. 3 shows the MAAAQ (mobility, authentication, authorization, accounting, and QoS) entity is built on top of

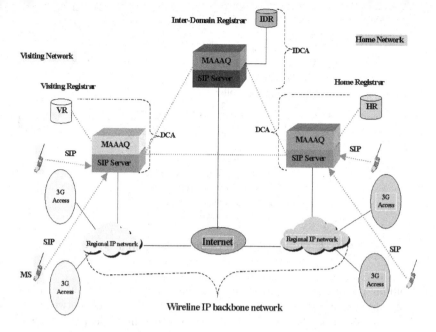

Fig. 3. ITSUMO's long term signaling architecture

a SIP-based DCA (and/or IDCA) system, and uses the location and signaling services of SIP to support roaming users.

3 Mobility

The issue of mobility can be discussed by first considering its implementation in the telephone system, and then user and terminal mobility in the Internet.

3.1 Mobility in the Telephone System

In a telephony system, a fundamental requirement is to be able to locate a user in order to establish real-time communication. Conventional telephony does not provide a means of locating a user per-se, but rather assumes a relatively static association of a fixed terminal with each user, and relies on the caller to consult a directory to map user names to terminal numbers (addresses). The user being called can temporarily circumvent this scheme by enabling the call forwarding feature, allowing all calls addressed to a given terminal to be redirected to an alternate terminal. In most telephone systems it is also possible to get a so-called personal number - a telephone number associated with an individual user rather than a given terminal. The system maintains a dynamic mapping between the personal telephone number and the number of a specific telephone. This mapping is updated explicitly by the user to whom the number is assigned.

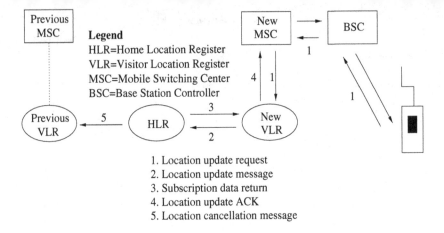

Fig. 4. Mobility in the Cellular Telephone System.

Cellular telephony uses the same basic model as conventional telephony except of course that terminals are mobile and allowed to roam between coverage areas (cells). To establish a call, the system must first locate the current location of the target terminal. This is achieved by accessing a database at the "home" location of the terminal, as determined by the terminal number. This database, usually called the Home Location Register (HLR), maintains an entry identifying the terminal's last recorded location. Each coverage area has its own database that maintains entries for all visiting terminals - usually known as the Visitor Location Register (VLR). Fig. 4 shows the steps involved in this terminal location process. As a terminal moves from one coverage area to another in a process known as handoff, an entry is created in the new VLR, the entry in the previous VLR is removed, and the HLR is updated to point at the new coverage area. Should the terminal be involved in a call when the move between coverage areas occurs, the handoff mechanism arranges for the voice connection to be transferred across to the new cell.

3.2 User Mobility in the Internet

In the Internet, the association between a user and a terminal number is not assumed to be static. Users expect to be able to access the Internet from any computer, and it is commonly the case that the terminal number (IP address) assigned to a given terminal is generated dynamically using the Dynamic Host Configuration Protocol (DHCP). Thus, to locate the address of the user's current terminal requires that a dynamic binding be maintained and consulted whenever one wishes to locate that user. Early VoIP products relied on the use of a special rendezvous service, such as an Internet Relay Chat (IRC) server, to allow users to browse the list of users online, and search for the current IP address associated with a user. Each time the VoIP client software executed it took care of the registration procedure on behalf of the user.

More recent approaches include the H.323 Gatekeeper, which provides a mapping between a user alias such as an email-address and the corresponding transport address. Use of an email address to identify a user provides a naming space which is convenient, scalable and can provide unique identification. H.323 client software can take care of registering the user with a Gatekeeper. The SIP Location Server provides the same function, but in addition to explicit registration, the server can also be configured to use other techniques to try to locate a user, including for example the user of the finger utility to probe a set of likely computers. The TOPS directory service allows user registration, but in addition supports the association of more than one terminal with a given user, and call processing logic to select amongst different terminals based on attributes such as caller identification, time-of-day and media types desired.

3.3   Internet Terminal Mobility

Host mobility refers to the situation where a terminal can be moved from one location to another. There are two cases to consider. Nomadic computing is when a terminal is used in many different locations, but in each case obtains a local IP address via DHCP. The salient point here is that any entries in user directory servers must be updated with the new address. The other model is mobile computing, where a terminal is allowed to retain its IP address no matter in which location it happens to be operating. This is the problem that the Mobile-IP protocol [12] solves.

Mobile-IP introduces the concept of a terminal having two IP addresses: a home IP address and a "care-of" IP address. The home IP address refers to the address the terminal is assigned in its home subnet. The care-of address is a temporary IP address assigned for use by the terminal while it is visiting another subnet, so it changes each time the terminal moves from one subnet to another. The basic idea is as follows. When a terminal moves outside its home subnet to another subnet it first acquires a local care-of address, and registers that address with a Home Agent back on its home subnet. Packets addressed to the terminal continue to arrive on the home subnet, are picked up by the home agent and forwarded to the terminal at its current care-of address (possibly via a Foreign Agent). This is shown in Fig. 5. When a terminal moves between subnets there is a handoff procedure that takes care of updating registration information at the Home Agent.

Several performance issues arise when considering the use of Mobile-IP for supporting VoIP. First there is the issue of end-to-end latency, which for voice telephony must be tightly constrained. In the basic proposal for Mobile-IP, all packets destined for a mobile host travel via the host's home subnet - resulting in a situation known as triangular routing. Clearly this can result in significant increases in the end-to-end delay for a VoIP call. A solution to triangular routing exists as a route optimization protocol, which allows so-called corresponding hosts to learn the current care-of address of the target host. This protocol is designed as an optimization and can result in latency which is highly variable.

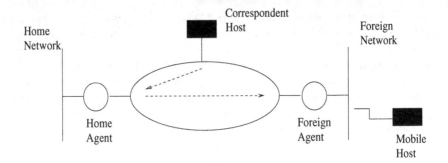

Fig. 5. Mobile-IP operation.

Another performance issue is the latency of handoff in Mobile-IP. Because of the need to interact with the Home Agent in order to perform a handoff, the latency involved can be significant. There have been several proposals designed to improve the performance of handoff such as [2]. This proposal introduces a hierarchy which allows several coverage areas to be managed as a group, within which host movement is invisible to the Home Agent and correspondent hosts. Related to that is the Cellular IP work [3] which suggests the use of location management techniques similar to those used in the cellular telephone system for providing a scalable mobility management solution with Mobile-IP. Finally, [13] proposes using SIP to provide low-latency mobility for terminals involved in multimedia sessions. In that solution SIP signaling messages are re-routed as usual based on the operation of Mobile-IP, but the media streams are re-routed at the application layer by having SIP renegotiate the various streams to use the care-of address.

4   Service Quality

In addition to mobility-related performance issues as discussed above, VoIP communication involving mobile terminals requires two key service quality issues to be addressed: the ability to schedule voice and non-real time data on a wireless access medium, and protocols to provide resource reservations in situations where one or more terminals can be mobile.

4.1   Medium Access

The issue of providing scheduled medium access has of course received considerable interest in the research literature, both for wireline and wireless networks. Popular wireless local area networks have followed a contention-based access model comparable to that of an Ethernet network, making them generally unsuitable for multimedia traffic. For example, the IEEE 802.11 standard for wireless LANs implements carrier sense multiple access (CSMA) as its default mode of operation, known as the distributed coordination function (DCF). Just as with Ethernet a consequence of this approach is an inability to predict access latencies, since they are dependent on the current offered load. The 802.11 standard also provides an optional mode of operation called the point coordination function (PCF) which introduces an access controller for each coverage area. The access controller implements a contention-free access period in addition to the contention-based period of the default mode. In the contention-free period the access controller polls those hosts which have indicated a desire to have predictable access, giving them a regular opportunity to access the medium.

A different solution is exemplified by the Wireless Integrated Services Platform (WISP) [4]. WISP supports a mix of constant bit rate voice and best-effort data by employing a medium access protocol based on token passing. In essence, a controller in each coverage area runs an algorithm which, on request from a terminal, grants access to the medium by passing a token to the terminal. For voice traffic, the terminal exchanges messages with the controller to specify its bandwidth requirements and hence the rate at which it will require tokens, allowing the controller to perform admission control.

4.2 Resource Reservation

Multiple service classes in an IP network can be provided by the resource reservation protocol (RSVP). Based on the notion of a data flow, RSVP allows a sender to describe the traffic it plans to transmit, and receivers the class of service they desire. This information is used to perform admission control and establish flow state at each of the intermediate routers. In RSVP, this information is soft-state, i.e. it times out and is discarded if not regularly refreshed. A key problem that arises is how RSVP works in the presence of terminals that are mobile. In particular, having established a reservation, if either or both sender and receiver move to a different network then a new reservation is required. This has performance implications relating to the delay of establishing a new reservation. It also raises the possibility of a terminal moving into a coverage area in which the desired resources are not available.

An extension to RSVP has been proposed to deal with these issues. Known as Mobile-RSVP [1], the idea is to define a reservation as being either mobility independent or mobility dependent. For mobility independent service it is necessary to make spatial reservations, i.e. at each of the locations that a mobile terminal might visit during the lifetime of the flow. This set of locations must be specified by the subscriber as part of a mobility specification. The scheme distinguishes between two types of reservation: an active reservation to the current location of the terminal, and passive reservations to all others cells in the mobility specification. A passive reservation becomes an active reservation when the terminal enters the corresponding coverage area. Subscribers to the mobility dependent service make a reservation from the sender only to their current cell. To avoid low utilization of the network, the resources associated with passive reservations are allowed to be used by other flows as if they were unreserved.

5 Conclusions

The use of IP to transport voice is receiving significant attention in the telecommunications and data networking industries. Commonly known as Voice-over-IP (VoIP), this technology has the potential to allow user communication involving multiple media types, and between terminals that offer improved capabilities and user interfaces. Meanwhile there has been an explosive growth in the demand for wireless connectivity, raising the question of how VoIP can be extended to situations where users and terminals can be mobile. In this paper we have surveyed the technologies necessary to provide mobile VoIP. It is clear that mobility has serious implications for locating users in order to establish a call, and maintaining connectivity with appropriate QoS throughout a call, even in the presence of mobility. Mobility also has an impact on call signaling, especially in relation to having to adapt call characteristics and media destinations as a terminal moves. We have discussed these key issues and given direction towards possible solutions.

Designing a network to provide wireline VoIP alongside other services is in itself a very difficult task, and introduces a host of technical and operational problems. In a mobile VoIP system there are additional issues of tracking users that can roam between different autonomous systems, and providing seamless handoff between wireless access infrastructures that are owned and operated by different organizations. These remain complex issues that need to be resolved within the overall goal of providing multi-service networks where voice telephony is just one of a set of services all based on IP delivery to the end-terminal.

References

[1] BADRINATH, B., AND TALUKDAR, A. IPv6 + MOBILE-IP + MRSVP = Internet Cellular Phone? In *Proc. of IFIP Int. Workshop on Quality of Service* (May 1997), pp. 49–52.

[2] CACERES, R., AND PADMANABHAN, V. N. Fast and scalable wireless handoffs in supports of mobile Internet audio. *ACM/Baltzer Mobile Networks and Applications 3*, 4 (1998), 351–363.

[3] CAMPBELL, A., GOMEZ, J., KIM, S., WAN, C., AND VALKO, A. A cellular IP testbed demonstrator. In *Proc. of the 6th IEEE International Workshop on Mobile Multimedia Communications (MOMUC)* (1999), pp. 145–148.

[4] GOEL, S., MISHRA, P., SARAN, H., AND SREENAN, C. Design and evaluation of a platform for mobile packet telephony. In *Proc. of IEEE International Workshop on Network and Operating System Support for Digital Audio and Video (NOSSDAV)* (July 1998), pp. 71–81.

[5] GOYAL, P., GREENBERG, A., KALMANEK, C. R., MARSHALL, W. T., MISHRA, P., NORTZ, D., AND RAMAKRISHNAN, K. K. Integration of call signaling and resource management for IP telephony. *IEEE Network 13*, 3 (May/June 1999), 24–32.

[6] HANDLEY, M., SCHULZRINNE, H., SCHOOLER, E., AND ROSENBERG, J. SIP: session initiation protocol. IETF RFC 2543, Mar. 1999.

[7] ITSUMO GROUP. Benchmarking of ITSUMO's all IP wireless architecture. Mobile Wireless Internet Forum (http://www.mwif.org/) <mwif2000.028.0>, Jan. 2000.

[8] ITU-T REC. H.323. Packet-based multimedia communications systems, Oct. 1997.

[9] KALMANEK, C., KAPLAN, A., MARSHALL, W., MISHRA, P., ONUFRYK, P., RAMAKRISHNAN, K., AND SREENAN, C. TOPS: an architecture for telephony over packet networks. *IEEE Journal on Selected Areas in Communications 17*, 1 (Jan. 1999), 91–108.

[10] KANTER, T., OLROG, C., AND JR, G. Q. M. VoIP for wireless and mobile multimedia applications. In *Proc. of the 1999 Personal Computing and Communications Workshop* (Nov. 1999), pp. 141–144.

[11] LIAO, W. Mobile Internet telephony: mobile extensions to H.323. In *Proc. of IEEE INFOCOM* (New York, NY, Mar. 1999), pp. 12–19.

[12] PERKINS, C. IP mobility support. IETF RFC 2002, Oct. 1996.

[13] WEDLUND, E., AND SCHULZRINNE, H. Mobility support using SIP. In *Proc. of 2nd ACM International Workshop on Wireless Mobile Multimedia* (1999).

Smart Delivery of Multimedia Content for Wireless Applications

Theo Kanter[1], Per Lindtorp[2], Christian Olrog[3], and Gerald Q. Maguire Jr.[4]

{[1] Theo.Kanter [2] Per.Lindtorp [3] Christian.Olrog }@era.ericsson.se,
Ericsson Radio Systems AB, SE-164 80, Stockholm, Sweden.
[4] maguire@it.kth.se,
TeleInformatics, KTH, SE-164 40, Stockholm, Sweden.

Abstract. Packet-oriented access to cellular networks enables us to deliver multimedia content to mobile users. As cellular networks will continue to deliver circuit switched voice for some time to come, care must be taken to avoid interference between these delivery mechanisms, while maximizing the range of services and the number of users. Smart delivery of multimedia content involving agents running in the mobile, the base station and the content provider allows us to dynamically adapt the application and network behavior to each other in order to meet the criteria for specific applications. In particular, this paper examines the delivery of streaming media and interactive voice as Voice over IP (VoIP) to mobile users. Our conclusion is that this, in combination with the dynamic adaptive properties as introduced by the agents, enables us to transfer voice entirely IP over wireless links, thereby freeing further resources for the new applications that we refer to in this paper.

1   Introduction

Presently, the telecom and datacom industries are converging in different ways. With respect to mobile telephony with GSM, new devices are appearing on the market that integrate data with the telephony voice service in new ways. So-called Smart Phones either include the functionality of an organizer, or can connect wirelessly to an external Personal Data Assistant (PDA) and integrate the functionality of the organizer for smart dialing and messaging with the application that is running in the handset. The Wireless Application Protocol (WAP) is intended to move the point of integration of these services into the cellular access network. WAP-gateways can be used to adapt and convert Internet information, so that the mobile terminal can be used for interacting with a wider range of *network*-centric services (e.g. electronic payment, subscription to information services, unified messaging, etc.). However, WAP is neither intended nor well suited to transport multimedia content, but rather was targeted at simply extending GSM networks with data services. Only through WAP-gateways can these services be connected to services on the Internet. Therefore, WAP excludes mobile users from directly interacting with Internet content. On the other hand, simple low-bandwidth GPRS (General Packet Radio Service) is introduced in GSM-networks, which will provide direct Internet access to mobile users. GPRS enables the development of multimedia applications for the mobile

C.G. Omidyar (Ed.): MWCN 2000, LNCS 1818, pp. 70-81, 2000.

device. These applications can integrate content that resides on the Internet. EDGE (Enhanced Data-rate for GSM Evolution), the successor of GPRS, increases the bit-rate and thereby further relaxes the requirements on the mobile applications and Internet content, thus bring even more new multimedia applications to mobile devices. Mobile devices are now able to perform significant computations based on events from various input devices and/or information sources. These events provide information about the user's context and the conditions, which the link is facing. Therefore, applications in mobile devices and nodes in the network that co-operate to deliver multimedia services to users can adapt their mode of communication dynamically based on such events. In such cases, the requirements for the delivery of multimedia IP-content over a link using a wireless access networks are even further relaxed.

2   Problem Statement

The important question is therefore: how cleverly can we dynamically shape the applications and Internet content, in order to maximize the delivery of multimedia content to mobile users? Multimedia applications put wide-ranging requirements on links. However, the quality of service (QoS) requirements involved can be categorized by several parameters, which differ in importance for the different type of services (and applications). We examine these parameters below.

2.1   Latency

Latency is important for isochronous services, such as voice (e.g., interactive speech) where delays up to 250 milliseconds are perceived as acceptable. Beyond 500 milliseconds the behavior of users gradually adapts itself to the increase in delay and the receiving party usually waits for a ready signal before starting to send. Latency may be due to network-related delays, and buffering in either the application or the device. Latency is not critical in streaming applications, such as Internet radio and other applications that playback multimedia content (e.g. MP3-files) as long as the user is assured that the content is going to be received within a bounded maximum delay.

2.2   Robustness

Latency is also important for interactive network games, such as Quake and Unreal, but in this case delay is not correlated to congestion problems. On the other hand, it is important that the packets arrive, otherwise synchronization problems will occur (these are not critical as continuous updates of players' locations are sent). This is normally handled by TCP or by UDP and an application-specific handshake protocol. However, the link level may also provide this service. The interleaving that GPRS does increases the probability that a packet is *not* lost, but on the other hand the user pays a penalty in increased latency, even if the packet is **not** lost.

Ericsson's GPRS Application Alliance has tested the currently most popular network game, Unreal Tournament, with a GPRS simulator with good results [16]. Using only one timeslot, the latency (about 300 - 800 msec) at 26 kbps incurred by GPRS interleaving does somewhat adversely affect players' performance and appreciation of this application, but not in a critical way.

2.3   Speech Quality

The coding/encoding algorithms (codecs) dictate the upper bound of the perceived QoS of multimedia streams (such as speech, video or audio). The lower bound is dictated by the percentage of link packet-loss that the codec is able to tolerate before its performance suffers severely. Modern codecs (e.g., Voxware RT-24) can tolerate up to 30-40% packet loss, with no additional latency.

2.4   Requirements

In order to assure that a certain application is feasible we must show that we can meet the requirements in a satisfactory manner at all times and in a scalable way. Two services of particular interest that will be studied further in this paper are:
1. delivery of streamed audio (e.g., MP3-files, Internet radio stations) and
2. interactive voice - e.g. Internet Telephony

The question is how we should use the functionality in: the mobile device, the wireless link, and the network, in order to provide scalable services and applications, such that the number of users is maximized. When designing our applications, we can use the knowledge that applications have of specific end-user requirements to our advantage, specifically by using strategies that dynamically shape the applications and their use of Internet content. This can be done by carefully dynamically adapting the mobile device, the wireless link, and the network, for each application — such that we can assure that these dynamic adaptations will work in an optimal way. This is not to say that the network access needs to be application dependent. We assume that access to the network is based on IP over the wireless link. Using this IP-access, the application is able to negotiate for its resources. The adaptation software running in the mobile device utilizes local APIs.

A recent paper [1] shows that we can deliver some multimedia content as background IP-traffic over GPRS. In this scheme, special care must be taken to avoid interference with switched voice services – this results in a roughly 30% under-utilization, so as not to hurt the channel planning of the network operators. A successful strategy, which dynamically shapes the application's demands for Internet content, must address a number of issues:
1. We should avoid situations where we would require unnecessary over-provisioning of bandwidth or other network resources in order for the applications to work.
2. We should maximize the number of users that will be able to use the services in a mobile environment.
3. This will make the applications feasible at an earlier point in time, i.e. *before* network resources are further developed.

4. Such a strategy will save the network provider cost by avoiding investments in infrastructure that might mean *unnecessary* over-provisioning, while providing revenue from these new applications and services.

These issues are particularly important regarding IP-access to wireless networks, where a common assumption has been that we must wait until EDGE or W-CDMA is fully deployed before we can start using these new applications.

3   Proposal

We propose to use agents to represent the different entities in the mobile device, wireless link, and network. The advantage of such an approach is that the agent can behave *intelligently* on a local level. For instance, the agent can transform image content from color to black and white in order to reduce the data that needs to be transmitted, based on the device characteristics *or* the available bandwidth at a given price. In addition, the agents can act intelligently *in concert*, when a certain application demands resources, they can adjust their behavior depending on a *non-local context* (e.g., routing content to (storage) locations, or to where there *are* currently wireless connections with high-bit rates available). Agents may even incorporate machine learning mechanisms to improve their performance with respect to QoS over time.

A common objection to the usage of agents is that solutions involving agents require the global adoption of an agent-specific discovery and negotiation schema to make it work successfully. However, [5,6] show how we can avoid this pitfall by relying on a general signaling protocol (e.g., the Session Initiation Protocol – SIP [10]) for the location of resources and use it to set up a session for the entities to do agent negotiations when applicable.

In the following section, we will show how agents can be applied to the mobile device, wireless link, and network, to optimize the QoS parameters (that were mentioned in section 0) for the application.

3.1   Mobile Device

The application that runs on the mobile device is modeled as an agent. This software is able to use the local APIs to control the speech codecs, length of the sound buffers, choose between available service classes over the wireless link, and even choose between different wireless links.

Furthermore, in certain situations the application may benefit from roaming to another network node. Since the sending party expects a dialog with the receiving party to continue, even if the receiving party goes off-line, the delivery of content is delayed until the receiving party goes on-line again. Utilizing such knowledge about the purpose of the communication can be important in creating intelligent push-services.

In addition, an interesting side point is that advance knowledge of transmissions allows the mobile device to go off-line and hence on standby for longer periods of time, without harming the application, but with dramatic improvement in battery life.

3.2   Wireless Link

Adaptation of header compression profiles [11] and lower-level behavior in the wireless link (e.g., coding, etc.) should not be directly visible to the application. On the other hand, we could introduce an entity in the access network, modeled as an agent, which acting as a proxy will select a suitable adaptation on behalf of the application.

3.3   Network

In a combined GSM and GPRS base station we may include a content-management agent to monitor unused link frames and keep track of which channels are used for switched voice and which channels are used for packet data. This information can be used to increase the utilization of the available bandwidth for IP-connectivity as compared to the relatively static division planned for such base stations.

Furthermore, we can co-locate a SIP-server with the base station. The role of the SIP-server is to locate the user and certain application resources that (as proposed above) should be modeled as agents. Negotiations of sessions are done using SIP, directly between these entities, without involving a SIP-server.

This way, the content-management agent not only helps the base station to increase its throughput of packet data but can also negotiate with the receiver and sender concerning their needs versus available capacity. This provides a basis for developing strategies to adapt the communication in such a way that it fits the end-user's requirements in an optimal way.

The following sections describe two key multimedia applications and illustrate how the proposed functionality can be used to achieve our goals.

4   Streamed Audio

Streaming Audio as broadcasted by Internet radio stations, is not sensitive to delays. In fact, seconds or even minutes worth of buffering can be done. The critical part is to gain acceptance from the user for these types of delays. Recently, the downloading and playback of stored music (e.g., MP3) has become very popular. A forthcoming publication [1] shows how stored music can be forwarded to mobile devices as background traffic in GPRS-enhanced GSM-networks. In addition, user audio (e.g., from dictations or classes with hyperlinks to electronic notes) could be uplinked for later playback. With this type of application, it is understood by the end-user that downloading may take considerable time, but this can be acceptable as long as the download time (typically at night or during the workday) does not exceed the period of time between use.

4.1   Gross vs. Net Content

An interesting calculation is how many unique bits does a radio station produce per day (eliminating the redundant replays of songs, ads, etc.). Now consider an ensemble of radio stations which all play many of the same songs – but have different ads, announcements, and play the songs in a different order — how many unique bits does the ensemble generate? A radio station sending an audio stream at 8 kbps generates roughly 700 Mb during 24 hours. A typical song lasts three minutes. Typically commercials occur four times per hour and last 3 minutes, indicating the station can play 16 songs per hour. This provides room for 384 unique songs but radio station profiling, popularity of certain tunes, and marketing of new music, demand that certain songs appear much more often. Assume, the top ten is played once every two hours, and the next twenty half this rate, and the next ten again at half this rate, hence there will be only 3 random songs per hour. This indicates that the station transmits at best (102 songs * 3 minutes * 8 kbps =) 146 Mb of unique bits. As there are very many radio stations playing popular music network providers can save a lot of bandwidth in the backbone by either coordinating transmissions of identical content or pre-caching content in geographically distributed cells (see further section 0) and perhaps only transmit content identifiers.

4.2   Content Distribution

The agent in the content-server, the base station and the client can co-operate to adapt the communication in different ways. For instance, the agent in the base station may deduce that several end-users have subscribed to the same content and keep local copies in order to reduce traffic in the backbone. It can also multicast this content when possible. Furthermore, this agent can instruct the base station control software to fill unused frames with data, and also predict unused channels for potentially use of delayed the transmission of content. This requires that client agents running in the mobile should be able to deal with paused or even interrupted transmissions, but these are known and solved issues to many FTP-and other clients today.

Based on communication between the agents in the mobile and in the base station, the agent in the base station can note that some of the content does not have to be sent as all the mobiles in the call which would get it - already have this content in their cache.

In light of this, multicasting "ftp" downloads [17], which allow "data holes", should be investigated. The holes can be filled asynchronous to the main download. This way, a user choosing to download an MP3, which is already being downloaded (say halfway) by another user, can hook in to the existing downstream and fetch the rest separately. The actual delivery may even be delayed intentionally until a given number of subscribers are online, or a timeout is exceeded, introducing a notion of content "launch".

4.3   Caching

We may also think in terms of content delivery networks, for situations where radio resources exist, but the GPRS backbone does not have spare bandwidth. This calls for:
1. Web-server replication (geographically separated, co-located with base stations).
2. Reverse (transparent) cached proxying, possibly implementing hierarchical caches.

4.4   Buffering

A calculation of the necessary storage capacity of large buffers in the mobile device shows that you could go for long periods of time without having more than a very low data-rate high-latency background service. For instance 64 MB holds 7 hours of internet radio quality audio [1] – so at the ~32kbps, which is available during the peak of the day (if you could use all of it) — this amount of memory could be filled in ~16000 seconds (i.e., roughly 4.4 hours).

The capacity of a macrocell (as configured in [1]) can only supply about 64MB of *total transfer* during the peak of the day (unless you want to exceed the 2% call blocking probability). This means you either have to:

1. allocate more capacity to this traffic
2. utilize the unused individual frames within the on-going calls
3. download large portions of the content in hotspots (where you have more available capacity – either because of fewer demands in this cell or because the cell has a higher data rate)
4. download large portions of the content in off peak periods
5. have much larger buffers in the device - so you can pre-load even more content

Dimensioning of buffers is important, it allows us to deploy a Mobile Audio Distribution (MAD) [1] service, where the delay is acceptable as long as you experience continuous audio. It is the user experience or perception, which is important.

Based on the results in section 0, each Internet radio station would send 18.25 Mb/day, assuming some day-to-day coherence the actual amount is *even lower*. This means that during the peak of the day has sufficient spare capacity to support more than three such stations even if these stations had to transmit all their content only during the peak voice hours! All this while using ~32kbps background capacity.

4.5   Agents

Different strategies are possible where companies are able to do directed ads, potentially location or context-aware, paying for the extra cost that this requires in order to have shorter delivery times. Besides adding capacity to either the access network or the device, options 2-3 in section 0 are really the interesting ones where agents are able to help out.

Media files may be streamed from mobile devices to the agent on the access point for subsequent distribution to anyone who desires to listen in. Thus, with GPRS a reporter can collect interviews and broadcast them using at most one channel worth of resources, while simultaneously supporting people listening to MP3-based "stations" on their way home from work. An interesting aspect is when this agent finds superfluous capacity it attempts to pre-order content and trickle it down the wireless link for storage in the mobile.

Furthermore, radio stations often have automated programs, so a network provider could strike a deal with a radio station by buying access to these automatic programming schedules and just multicast the content interspersed with the advertisements to the users at the far end of the network. Exploiting advance knowledge of content programming would save network operators a significant investment while offering radio stations and content providers knowledge about what the user likes, thus offering benefits to all parties!

An important goal for research in the field of mobile computing and communication is to build so-called 'unconscious' services: automatic services that do not require user intervention to execute because the components involved are able to make decisions themselves [3,4,5]. An important reason for the use of agents is that they enable us to build such unconscious services. For instance: entering our future homes (where we will have high-bandwidth wireless connectivity between devices) our mobile device will connect to the networked stereo components (where the same information is pre-stored on the server) and hand over the on-going multimedia interaction with the user to it. Before going on a timed standby to save its batteries, the mobile device pre-orders the delivery of a copy of new multimedia content via the server to a content provider, according to the user's current preferences. Agents represent the components mentioned here. These agents only have to demonstrate reasonable local behavior in order to offer a very useful service in concert. The service relieves the user of consciously having to reload the mobile device with fresh content according to the current preferences.

5   Interactive Voice

The premise that voice will continue to be the most profitable application, which is upheld by many vendors and operators of cellular networks, is false. Voice can be delivered over IP access [9,10] over GPRS [2] even at bit-rates that will be available in the first generation of GPRS. VoIP over wireless can be delivered at 90% efficiency (of the used radio spectrum) as compared to switched voice), when smart header compression is applied [11,12]. Previously, we have shown that 1.2 kbps is all that it takes to deliver voice over IP over wireless [2], which is ~10% of a single GSM voice channel. In further support of the feasibility of VoIP over wireless, DiffServ is a reasonable way of guaranteeing bandwidth for most real-time sensitive applications [1,18] – see further section 0.

Then, we may ask: how long will the statement remain valid that much of the capacity will be used by switched voice, thus relegating anything else to be delivered in the "spare capacity". Delivering switched voice incurs a high investment and maintenance cost for network operators. For wired access, where there is plenty of bandwidth, the cost imbalance has already overthrown the model of switched voice. The "spare capacity" model will remain valid only until the use of VoIP in cellular networks increases sufficient to trigger a wholesale transition to packet-oriented wireless links. But with the experiment that 1.2 kbps is all that it takes to deliver voice [2], this transition cannot be too long after GPRS is introduced. At this average rate you can support at least 26 simultaneous additional VoIP calls in a macro-cell when during busy hours ~32 kbps of spare capacity is available. Assuming business calling patterns this is sufficient to support a population of hundreds of mobile voice users in the area of the cell. Naturally, there will be a QoS reduction caused by the lower speech quality, but end-users will most likely accept this if it means that voice is delivered essentially free in conjunction with other services.

5.1   QoS Considerations

Statically assigned integrated services should be avoided as much as possible, as it implies switched circuit connections over packet networks, whereas differentiated services are crucial to allow e.g. small VoIP packets to cut through the router's FTP and HTTP queues.

Adaptive "loose" integrated services may be necessary, e.g. monitoring the frequency and recurrence of certain high priority packets, capable of preventing issuance of large size packets which would seriously interfere with the probable next high priority packet.

5.2   Base Station

In a recent paper [2] we proposed to use IP directly over the wireless link and thus replace the GSM base station with a router with a radio, thus mobile users will be able to use VoIP for the delivery of interactive voice (see also [7,8]). As compared to the scenario in [1] where switched voice is prioritized, this would provide for much more flexible use of the available bandwidth. The following scenario also holds for the combined GSM- and GPRS-base station, but it is clear that the headroom for dynamic intelligent behavior of the access point diminishes as more channels are allocated/dedicated for switched voice traffic.

It is obvious that such a base station with a content-management agent can adapt the transmission of content in a fair profitable way. As above, the access point can be co-located with a SIP-server.

5.3   Agents

With respect to VoIP, similar scenarios as mentioned in section 0 apply. However, as VoIP puts other requirements on the mobile device, wireless link, and network, agents can help to further adapt the characteristics of these to the current communication context. For instance, if the available bit rate drops, the agent in the mobile device can renegotiate the codec that is to be used, it may also contact the agent in the base station to temporarily halt the transmission of other content that is not sensitive to delays. In case of packet loss, the agent can suggest a change in header compression profiles [11]. The agent in the mobile device could also negotiate with agents located in other base stations for better conditions and more favorable price per bandwidth and cause the communication to be handed over to one of them. Other scenarios may take in to consideration the context of the user (e.g., when the mobile device contains sensors), affecting the content of the voice content. For instance, the content provider could transmit directional audio for augmented reality applications.

6   Conclusions

In telephony applications the load per subscriber during the peak hour corresponds to 3 minutes / 60 minutes = 0.05 Erlang, where a blocking probability of 1% can be tolerated. Assuming we can transmit 26 simultaneously VoIP sessions (see section 0), means that we can serve ~500 voice users in a cell just using spare capacity during peak hours. On the other hand we can make reasonable assumptions about user behavior. If users are talking they will be using other services less (e.g., browsing the web or downloading MP3 files). This means a user agent running with the application in the mobile device can mediate between the two applications that were mentioned in this paper. VoIP applications could convey information on speech activity, similar to 'push to talk' by monitoring input to the speech buffers and thereby signal to the multimedia download application to temporarily pause transmissions [14]. This can also be done by statistical prediction and a 'back off' mechanism.

In addition, for streamed audio applications a much higher blocking probability can be tolerated as long as we can show that we have a throughput that ensures that the content is delivered within a bounded maximum period of time (see also section 0). Above we showed that an agent in the mobile device can mediate on a local level between VoIP and streamed audio download applications. Similarly, the content-management agent can make decisions about the fair distribution of capacity among users of these two applications within the scope of a GPRS/GSM base station through a dialog with the SIP-redirect server. Thereby, the agent has knowledge of ongoing sessions and as mentioned above, can predict traffic and thus assist in planning the transmission of additional Internet content.

The advantage with a agent in the base station is that it can observe the link and utilize the pre-stored bits (MP3, etc.) it has to fill voids in the utilization link - you can have high link utilization but still let voice packets through on time. The same argument regarding monitoring of communication content can be applied to the agent in the mobile device. Speaker dependent speech recognition can be applied using the increased computing capability that is available in mobile devices. A new device first learns "his master's voice". Therefore, voice can not only be surpressed to the point that we send ASCII text "annotated" with voice inflections (consequently causing a further reduction in consumed bandwidth for voice). An agent monitoring content in the mobile device will also be able to alter the mode of communication based upon what is actually being said, which is very useful in applications.

The implications to wireless business models are far-reaching. With GPRS, we have seen that customers get direct access to Internet content. As a consequence, third parties (meaning basically anybody — from private persons, local organizations, to radio- and television broadcasting companies) can provide content to end-users. This means that the network operator has no unique role, unlike the case with WAP gateways. Instead the network operator must seek a new role besides providing competitive price per unit of bandwidth for access to Internet, by offering intelligent support for smart delivery of multimedia content to mobile users. This means providing added value to end-users. At the same time, hosting of Internet content and smart transmission to mobile users creates a business opportunity for network operators to offer these services to content providers. Network operators offering agent hosting can also sell their end-users as a potential audience to those who want to send commercials.

In this paper, we proposed to use agents to represent the different entities in the mobile device, wireless link, and network (that can either behave *intelligently* on a local level *or* act intelligently *in concert*, based on a *non-local context*). By numerous examples we have demonstrated how this approach can be successfully applied to dynamically shape the applications and Internet content, in order to maximize the delivery of multimedia content to mobile users, by taking into account the context of the users and the conditions of the communication. In addition, the transferal of switched voice to VoIP further frees resources for the dynamic shaping of applications and our conclusion is that this transition will take place not too long after GPRS is introduced.

In conclusion, we have demonstrated by our approach of smart delivery of multimedia content in wireless that we can:

1. avoid situations where we would require unnecessary over-provisioning of bandwidth or other network resources in order for the applications to work.
2. maximize the number of users that will be able to use the services in a mobile environment.
3. make the applications feasible at an earlier point in time, i.e. before network resources are further developed, and
4. that such a strategy will save the network provider cost by avoiding investments in infrastructure that might mean unnecessary over-provisioning, while providing revenue from these new applications and services.

7　Future Work

Future work will look at the specifics of the interaction between the SIP-server, the content management agent in the base station, the agent located at the content provider, and the agent in the mobile device. Furthermore we will study how the agent in the mobile device can interact with communication resources, in particular those governing VoIP QoS, e.g. DiffServ and mobility. Another issue that needs to be addressed concerns how the content-management agent interacts with the radio resources in the base station.

8　References

1　P. Lindtorp, "Utilizing Spare Capacity in Radio Access Networks", Master Thesis, School of Electrical Engineering and Information Technology, Royal Institute of Technology, Sweden, December 1999.

2　T. Kanter, C. Olrog, G. Maguire, "VoIP over Wireless for Mobile Multimedia Applications", Personal Computing and Communication Workshop, November 1999.

3　T. Kanter, C. Frisk, H. Gustafsson – "Context-Aware Personal Communication for Teleliving", Personal Technologies (vol. 2 issue 4, 1998: p. 255 - 261).

4　T. Kanter and H. Gustafsson, "VoIP in Context-Aware Communication Spaces", Proceedings of International Symposium on Handheld and Ubiquitous Computing (HUC 99), Oct. 1999.

5　T. Kanter and H. Gustafsson, "Active Context Memory for Service Instantiation in Mobile Computing Applications," Proceedings of the Sixth IEEE International Workshop on Mobile Multimedia Communications (MoMuC'99), Nov. 1999, p. 179-183.

6   T. Kanter, "Adaptive Personal Mobile Communication", forthcoming Licentiate Thesis, Dept. of Teleinformatics, Royal Institute of Technology, Sweden.
7   C. Olrog – "GSM SoftModem on Linux, a Direct Radio Link Protocol Interface", Master Thesis, Dept. of TeleInformatics, Royal Institute of Technology, Sweden, April 1999.
8   T. Turletti, H. Bentzen and D.L. Tennenhouse, ``Towards the Software Realization of a GSM Base Station'', IEEE/JSAC, Special Issue on Software Radios, Vol. 17, No. 4, pp. 603-612, April, 1999.
9   ITU-T Recommendation H.323 v2 (1998) Packet Based Multimedia Communication Systems.
10   M. Handley, H. Schulzrinne, E. Schooler, J. Rosenberg - RFC 2543 on SIP: Session Initiation Protocol, IETF/Network Working Group – March 1999.
11   L-E. Jonsson, M. Degermark, H. Hannu, K. Svanbro, "RObust Checksum-based header COmpression" (ROCCO), IETF Network Working Group Draft, September 1999.
12   M. Engan, S. Casner, C. Bormann - RFC 2509 on IP Header Compression over PPP, IETF/Network Working Group, February 1999.
13   J. Mitola III, Cognitive Radio, Licentiate Thesis, Dept. of Teleinformatics, Royal Institute of Technology, Sweden, Sept. 1999.
14   J. Mitola III, "Cognitive Radio for Flexible Mobile Multimedia Communications," Proceedings of the Sixth IEEE International Workshop on Mobile Multimedia Communications (MoMuC'99), Nov. 1999, p. 3-10.
15   V. G. Bose, "The Impact of Software Radio on Wireless Networking," Mobile Computing and Communications Review, Volume 3, No. 1, January 1999.
16   News articles on www.planetunreal.com, "UT Over Wireless" and "UT Over GPRS", 11/19/99.
17   J. Ioannidis, G. Q. Maguire Jr., "The Coherent File Distribution Protocol," RCF 1235, IETF Network Working Group.
18   S. Black, D. Black, M. Carlson, E. Davies, Z. Wang, W. Weiss, "An Architecture for Differentiated Services," RCF 2475, IETF Network Working Group

IPv6 : The Solution for Future Universal Networks

Sathya Rao

Telscom AG, Sandrainstr. 17
3007 Bern, Switzerland

Abstract. The communication networks and services are changing rapidly. The conventional circuit and packet switched networks are being replaced by next generation networks, primarily based on Internet Protocol. The rapid growth of web based services has lead to the explosive growth of the internet. However, the current internet protocol (IPv4), which is the backbone of transmission control protocol (TCP/IP) networking, is rapidly becoming obsolete, with the inherent problems related with limited address space, security and QoS features. The new protocol IPv6 has been developed to overcome all these problems and to provide solutions for the next generation networks. This paper addresses the features of IPv6 Protocol, the status of standardisation, and various activities around the world.

1   Introduction

Europe's leadership in Internet technology and provision of user access should be based on an offering with unlimited address space, quality and security, properties the current Internet does not cater for. Europe should foster a unique leadership strategy in promoting the next generation networks based on the new Internet Protocol version 6 (IPv6) protocol in order to promote pan-European E-commerce, offering customer protection and benefits in terms of security and quality as services converge to run over IP. Such a Euro-IPv6 network will place Europe into a position of strength in comparison to the US with respect to New Internet technology.

The deployment of IPv6 requires a good spread of diverse technologies and the support of national Internet Service Providers across the whole European community. Expertise in these new technologies, which overcome the limitations of the current IPv4-based Internet, cannot be found in just a couple of European countries. The skills required lie in the areas of seamless deployment of IPv6 into a large existing IPv4 base, and provision of quality of service and security at host and gateway/router level.

The Internet has doubled in size every year since 1988. There are over 44 million hosts on the Internet and an estimated 200 million users world wide. By 2006, the Internet is likely to exceed the size of the global telephony network, should IP telephony have not replaced the existing telephony network by then. Moreover, tens of millions of Internet-enabled appliances will have joined.

C.G. Omidyar (Ed.): MWCN 2000, LNCS 1818, pp. 82-91, 2000.

The Telecom industry (manufacturers and operators) need to build strategies to cater for the mobile information society, deploying brand new products and services such as wireless Internet devices, Internet cell phones and personal digital assistants which will emerge to become the new telecommunications tool of the next decade. The mobile information society will need to deploy for this purpose IPv6 as a robust Internet foundation.

This strategy will get the European leadership entrenched in every aspect of the European industry in view of the explosion of the E-business and E-entertainment in Europe. It is estimated that commerce on the network (E-commerce) will reach somewhere between 1.7 T$ and 3.0T $ by 2003. That is only three years from now (but a long time in Internet years). Secure E-business and European privacy should be advocated and implemented at the network layer not just at an application layer. The security feature is built-into the IPv6 protocol to solve the issue, which is one of the major weakness of the current internet protocol.

2   Applications with IPv6 at Driving Seat

E-commerce will be a new driving force for new economies, creating new business sectors and new jobs across Europe. This new economy needs a robust platform to guarantee its success. The E-business and E-shopper should be able to gain the necessary confidence that they are doing business in a safe environment, which is not the case today. European E-commerce could back-fire in the mid to long term without adequate customer protection. E-commerce will also create the need for more address space and this new need is a healthy sign of the growth of the Internet in Europe, but then this growth has to be supported by guaranteed IP address space which is not available today.

The Internet is proving to be one of the most powerful amplifiers of speech ever invented. It offers a global megaphone for voices that might otherwise be heard only feebly, if at all. It invites and facilitates multiple points of view and dialogue in ways not possible through traditional, one-way mass media.

The Internet can facilitate democratic practices in unexpected ways. The proxy voting for stock shareholders is now commonly supported on the Internet. Perhaps we can find additional ways in which to simplify and expand the voting franchise in other domains, including the political, as access to the Internet increases.

The Internet is becoming the repository of all we have accomplished as a society. It is becoming a kind of disorganized Boswell of the human spirit. Shared databases on the Internet are acting to accelerate the pace of research progress, thanks to online access to commonly accessible repositories.

3   Ongoing Activities

The US government is funding the 6REN/6TAP testing native IPv6 as well as Internet2, a project that tests the impact of QoS and higher bandwidth on the current Internet. Internet2 has subscribed now to IPv6 as the result of their tests that higher bandwidth is not the only solution but a smarter packet is needed to achieve a better quality of service.

The Japanese government is funding the Wide Project which is a copy of the 6REN/6TAP initiative to take Japan into leadership in the New Internet. The Japanese government is pushing for active IETF involvement in order to secure RFC adoption or early RFC influence.

The IPv6 Forum has been formed recently to promote wide adoption of IPv6 specifications in developing next generation network products and services. The IPv6 Internet Initiative is a key milestone for a range of products and services under definition within the mobile information society platform. European Concepts based on GPRS, UMTS and 3G products and services depend dramatically on the deployment of IPv6. The convergence is an opportunity for the European switch manufacturers to take leadership into the New Internet and define Core Switch/Routers.

Within the European Union 5th framework 'Information Society Technologies' framework, a project named 6INIT has been started to promote the deployment of IPv6 networks and services, in collaboration with Japanese and Canadian partners.

Eurescom has initiated a project (P702 : Internet Protocol Version 6 - new opportunities for the European PNOs) to investigate the usage of IPv6 networks to replace the conventional networks for delivering conventional and future public services. Eurescom has also initiated a new project (P1009) to study the deployment and transition strategies for services on top of IPv6, e.g. Mobility, QoS support, Multicast Implementation, Network configuration and management.

4   Standards Status

4.1   The Internet Engineering Task Force and IPnG

The IETF (Internet Engineering Task Force) is very active in promoting IPv6 standards, through their Request for Comments (RFC) documents which are generally adopted as standards for implementation. IETF has constituted a special group IPnG (IP next generation) to promote IPv6 activity.

The current version (4) of the Internet Protocol (IPv4) uses node addresses that are allocated from a 32-bit space This 32 bit address space is further classsified to provision Class A, B and C ranges, which constituted network part of 8, 16 or 24 bit, with corresponding host part of 24, 16 or 8 bit, depending on the number of expected hosts on a given network. This led to inefficient use of the 2^{32} possible addresses, since many 'important' organisations automatically asked for class A or B addresses using up 2^{24} or 2^{16} addresses at each single assignment, even when they often only had several host computers, or had many subnets with several computers on each. A second problem was that addresses were rarely re-claimed after they were no longer in use.

The terms of reference for the working group is maintain all good features of current protocol specifications, and enhance the features to guarantee smooth transition to next generation networks. The IPv4 protocol is simple, binds multiple protocols, simple management, but limited with scalaibility in terms of address space, topological flexibility, Quality of service support, security, etc..

IPv6 is planned to support very high speed (Gbps), range of subnets, low information loss and will function independent of media (terrestrial, mobile, radio and satellite), provides auto configuration possibility, high security for business applications, application specific QoS, and multicast addressing facility. IPv6 will be also backward compatible to work with the current IPv4 protocol (through tunnelling mechanism).

4.2 IPv6 Features

The **next generation networks** based on IPv6 will provide:
- 128 bit wide address space to cover all possible appliances connectivity
- Differentiated Services in terms of quality (bandwidth guarantee and transit delays for real time flows).
- Security in terms of access point authentication, message integrity and privacy.
- Auto-configuration and reconfiguration capabilities allowing easy modification of network architectures.
- Management facilities allowing the setting up of on-demand services and providing ISPs with accounting capacities.
- Wide range of applications and services.
- Mobile host capabilities allowing provision of transparent access whatever the physical access used, supporting the evolving UMTS capabilities, will be the issue of co-operation between the mobile IP related projects (e.g. WINE).

4.3 Activities in the IETF

Within the IETF now, detailed work on IPv6 specification is persued. Changes to routing protocols, transport protocols (the pseudo-header checksum in TCP and UDP) and applications that reference IP addresses (particularly DNS, but also FTP) have been specified. More subtle wok in routing (beyond OSPF and RIP v6 changes) needs to be done, and more especially, the impact on RSVP and on multicast routing and Mobile IP routing as well as RTP/SDP and MPLS needs a lot of work to see what the real benefits may be.

The critical missing piece in IPv6 is a deployment plan that includes seamless interworking with IPv4, but provides clear benefits to a site to migrate. Three possible ways this may happen are: VPNs, Satellite IP (DBS) and large scale PDA/GSM mobile phone integration by a provider and vendor. The importance of the seamless interworking is because the Internet is now far too large to envisage event the switchover that occurred in going from previous NCP to IP in 1980; and that switchover was even painful then - with a few hundred hosts rather than tens of millions. Moreover, there is clear reluctance by commercial vendors to invest into extensive upgrade; it must be made worthwhile by the quality of the benefits provided.

4.4   Available Implementations, Products, and Services

We can divide implementations into host and router side code. In the router side, most of the major vendors have at least beta products for the basic IPv6, although its not clear if their routing protocols changes they still interwork with the legacy IPv4 networks. On the host side, major operating systems such as Windows 2000 will have IPv6 in, although to date, only the research part of Microsoft have released a Windows implementation. In a recent IPv6 Summit Microsoft announced their commitments to IPv6 and officially released the IPv6 protocol stack. Similarly CISCO also announced their commitment to the IPv6 networking, which provided early boost the next generation networks world. For Unix systems, there are releases for most major flavours, although many are very early code, and have a number of shortcomings. The best systems are the public domain offerings for FreeBSD and Linux, including DNS for IPv6 and other important infrastructural tools.

The implementations have many of the components that have been defined - but not all. For example, strong security is mandatory in IPv6; political considerations related to export controls have made it very difficult to have exportable implementations which meet the Standards. Moreover, some of the key components, like IP Multicast, Quality of Service facilities and Public Key Infrastructure are not yet fully standardised.

4.5   Address Space in Reality

In practice, the IPv4 address space has lasted longer than expected due to two technologies: Classless Inter-Domain Routing (CIDR), and Network Address Translation (NAT). This has reduced the urgency to move to IPv6.

CIDR is a generalisation of the class based address assignment that was originally devised for IPv4. Nowadays, the address assignment authorities (and they are devolved to regions of the Internet geographically now) assign IPv4 (and IPnG (or IPv6 as it is known to some)) addresses hierarchically, together with masks. The mask determines how many bits of the address are network and how many are host, and the mask can be different at different places in the network topology - this allows the address + mask to be treated like a variable length prefix. The forwarding decision that used to be made by routers, based on simple best-next-hop, is no longer a simple lookup; it consists of a longest-match procedure. Routing protocols no longer exchange lists of network numbers to build upon a network map, but now exchange addresses + masks, to allow this hierarchical address space management to work. A secondary, but very important side effect of this is that the routing tables can now be summarised; typically, a country might be assigned a short mask, and within the country, each region, longer masks. This allows each router nearer to each local region to contain small number of (even just 1 per interface) entries.

The questions of the technical and political feasibility of address-space deployment are relatively separate. One advantage, for example, of the larger address space is that mobile users can keep their same low-order addresses, while the mobile network operator controls only the high-order bits. However when there is a real conflict between technical and commercial pressures, the solution is less clear. It would be possible to carve out a complete set of numbers and address for IP-telephony; early attempts to do this in conjunction with the ITU telephony groups have failed so far. One suspects that this is partly because attempts to make IP numbering as universal and well-structured as the telephone numbering is seen as a very real threat by the PNOs to their telephone revenue.

NATs are another technique for containing the growth of address use, but are based in two other important requirements: to avoid having to renumber hosts, and to provide some network security. Many sites had assigned IP addresses in isolation from the Internet, and had used addresses already in use in the world-wide Internet. To avoid the problem of re-numbering all their hosts and routers, such sites developed a technology to translate ´on-the-fly` the addresses of source hosts within IP packets being forwarded through firewall routers. This was done only for hosts which wished to communicate with the ´outside world`, changing the addresses in packets to and from them, from the internal non-unique ones, to ones drawn (dynamically assigned) from a small pool of legitimate public addresses. This works well in typical large corporate networks as few hosts communicate with the outside world at any one time (principal of locality!). Often, the dynamic assignment in the firewall router was controlled by an application level authentication protocol (e.g., based on an RPC mechanism, or even just a telnet/login user-name + password to the firewall router). This meant that the NAT acts as a quite effective barrier to outside hosts accessing internal hosts (even if packets could get in by pure random luck, the responses would not get out...).

4.6 IPv6 Roadblocks

Many of the improvements promised for IPv6 have been specified in a simplified way for IPv4; the specifications are often less powerful than is possible with IPv6, but would provide enhanced services. It has turned out to be very difficult to organise even the IPv4 deployments of IPSEC, QoS and Mobile IP. This is partly from lack of motivation by the many smaller ISPs, but it is also because of the need to orchestrate the introduction of the services. There is often little advantage in introducing such services in an isolated part of the system. Moreover, in some cases the specifications are not really complete, or it is felt that their viability yet to be demonstrated by the R&D community. Here there is the serious problem that small-scale deployment is no feasibility demonstration; large-scale deployment is often beyond the means of the R&D community, and is not really supported by those providing the research networks.

It is probable that these improvements will come more with IPv6 than earlier, because the whole of the transition to IPv6 must be managed to some extent in any case. This will encourage the establishment of larger testbeds, at least by the bigger ISPs and PNOs. There are some substantial test-beds already; who are active on the 6Bone, which is a set of European sites who are experimenting with IPv6 in an encapsulated form. During the last summit NTT (from Japan) announced the world's first ISP to support IPv6.

5   Evolution Scenarios

The Internet engineering community is promoting a new version of the IPv6 as the answer to the address shortage predicted for the current Version 4. IPv6 offers enough addresses that every computer, cell phone and set-top box can be hooked up to the 'Net. However, migrating a large network to IPv6 is so difficult that few organizations have committed to it.

Theoretically, Version 4 could support up to 4.2 billion devices, but the allocation of those addresses has not been very efficient. An attempt has been made to increase the efficiency with interdomain routing and allocation rules that go along with it. But the side effect of those rules is the proliferation of network address translation [NAT] boxes, which take a single Internet address and multiplex it among a bunch of different devices. It's a fairly ugly process from an architectural point of view, although it turns out to be very effective, and a lot of people are relying on it. But because NAT intervenes at the IP address level, it has some consequences for end-to-end security and integrity of the traffic

The key transition objective is to allow IPv6 and IPv4 hosts to Interoperate. A second objective is to allow IPv6 hosts and routers to be deployed in the Internet in a highly diffuse and incremental fashion, with few interdependencies. A third objective is that the transition should be as easy as possible for end-users, system administrators, and network operators to understand and carry out.

Probably the most straightforward way to introduce IPv6-capable nodes is a dual stack approach, where IPv6 nodes also have a complete IPv4 implementation as well. Such a node, referred to as IPv6/IPv4 node in [RFC 1993], thus has the ability to send and receive both IPv4 and IPv6 packets. When interoperating with an IPv4 node, an IPv6/IPv4 node can use IPv4 packets; when interoperating with an IPv6 node, it can speak IPv6. IPv6/IPv4 nodes must have both IPv6 and IPv4 addresses. They must furthermore be able to determine whether another node is IPv6-capable or IPv4-only. This problem can be solved using the DNS, which can return an IPv6 address if the node name being resolved is IPv6 capable, or otherwise return an IPv4 address. Of course, if the node issuing the DNS request in only IPv4 capable, the DNS returns only an IPv4 address.

In the dual stack approach, if either the sender or the receiver is only IPv4-capable, IPv4 packets must be used. As a result, it is possible that two IPv6-capable nodes can end, in essence, sending IPv4 packets to each other. This is illustrated in Figure 1.

The target scenario is to have an IPv6 backbone network, which can also provide seamless interconnectivity with legacy IPv4 network. The typical network scenario with IPv6 backbone is shown in the Figure 2.

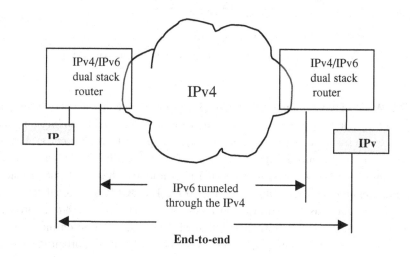

Figure1: A dual stack approach

6 Mobility and Internet

Europe is very strong in mobile networks deployment and usage. The internet access through mobile terminals and having an interactive data communication is a priority area in Europe. The fist such applications already being introduced based on WAP.

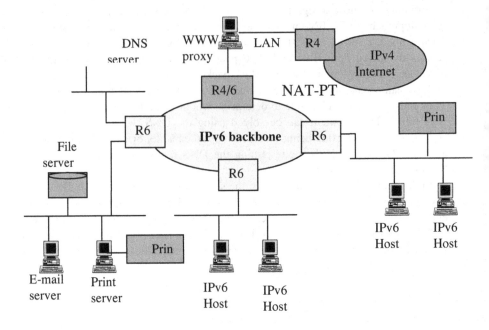

Figure 2: IPv4 and IPv6 network interworking

The WAP based communications are slow due to bit oriented non-efficient WAP protocol, though it provides the transition step for introducing mobile internet to the market. Third generation networks deployment plans to implement UMTS services are already in advanced stage of realisation. To progress the evolution and to enhance both mobility and internet features, the new initiative has been put in place called Third Generation Partnership Project (3GPP). The 3GPP is a global standardization initiative that was created just over a year ago, in December 1998, to produce technical specifications for Third Generation Mobile System based on the evolved GSM core networks and a new radio interface (UTRA). Major important steps have been achieved since then, such as the approval of Release '99 specifications in December 1999. The 3GPP work plan for the year 2000 includes Internet Protocol (IP) based communications. It is expected that as mobile phones gain access to Internet services, there will be an unprecedented growth in the demand of new Internet addresses as well as easier administration and tighter security. Convergence of Internet and Mobile Telecommunications move a step closer since IPv6 Forum has joined the 3GPP as a Market Representation Partner.

7   Conclusions

The specifications for a Next Generations Internet are largely complete from a technical viewpoint. There are still some loose ends, but they are not very significant. The large-scale deployment of many of the newer features over the current Internet is proving difficult, and will probably never happen on a large scale. Implementations of the basic feature sets are available in research prototypes, and starting to become available in commercial offerings. Implementations of advanced features are becoming available in research prototypes, but still require substantial experimentation and refinement. However, there are starting to be a number of substantial research networks on which the implementations are being deployed for R&D purposes. The general commitment to large-scale commercial deployment, and the time-scales over which this could be achieved, are still under discussion, though the recent announcements from the major vendors has brought the time scales to near short term plans.

Performance Measurement Methodologies and Quality of Service Evaluation in VoIP and Desktop Videoconferencing Networks

Henri TOBIET[1] and Pascal LORENZ[2]

[1] NMG TELECOMS – Network Management Group – Telecoms Solutions
20e, rue Salomon Grumbach, BP 2087, 68059 Mulhouse Cedex, France
h.tobiet@nmg.fr
[2] University of Colmar - IUT / GTR
34 rue du Grillenbreit - 68008 Colmar, France
lorenz@colmar.uha.fr

Abstract. The present paper relates to performance and quality of service evaluation for VoIP (Voice over IP) and desktop videoconferencing services in IP networks. Simulation-based performance measurements consist in the generation of performance statistics obtained by measurements realized by simulation on the subscriber interface. They include detailed measurement of call quality, call set-up quality and availability. The tests are accomplished by emission of non-disturbing additional test traffic. The real-time QoS monitoring is based on non-intrusive analysis of real call parameters. The advantage of this method is the exhaustiveness of the analyzed call. The QoS measurements are mainly service availability and call set-up quality.

1 Introduction

QoS defined in CCITT Recommendation E.800 may be considered as the generic definition reproduced below: *"The collective effect of service performance which determine the degree of satisfaction of a user of the service"*.

Quality of Service on the LAN represents a major challenge, not so much during the predictable processes of compressing the voice streams and splitting them into packets which are mathematically predictable, the real challenge is of sharing the connectionless transmission media with other users in a predictable and quantifiable way. As soon as voice/video traffic reaches the IP network, it must compete with electronic mail traffic, database applications and file transfers [13], [17], [18].

QoS evaluation methodologies are based on the previous studies performed in European Projects, such as QOSMIC (QoS Methodologies and tools for Integrated Communications). The resulting QoS will be dependent on the performance of the physical, AAL (ATM Adaptation Layer) and transport layers involved in the protocol stack implementation [1], [10].

Work is currently on-going in modelization of the TCP/IP protocol layer, in terms of performance aspects (delays, etc...). The resulting model will then be applied to

C.G. Omidyar (Ed.): MWCN 2000, LNCS 1818, pp. 92-107, 2000.
© Springer-Verlag Berlin Heidelberg 2000

convert network layer performance (ATM, AAL) to transport layer performance, according to the selected services: telephony over IP, file transfer over IP and videoconferencing over IP [12], [14].

1.1 QoS Characterization

The used TCP/IP model can be described as follows:

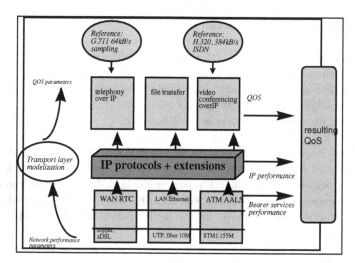

Fig. 1 : QoS protocol oriented approach

The main performance parameter to be measured is the Round Trip Time (RTT) but the performance evaluation will also concentrate on the following aspects: connections opening and closing mechanisms, data transfer mechanisms, addressing, parameters negotiation, congestion control and errors control.

Part of our studies concentrated on end-to-end QoS characterization. End-to-end QoS in a videoconferencing system is characterized under two broad headings:
- call set-up quality and
- call quality.

Call set-up quality is mainly characterized by the call set up time, i.e. the time elapsed from the end of the user interface command by the caller (keypad dialing, email alias typing, etc) to the receipt by the caller of a meaningful tone. ITU-T Recommendation E.600 provides more information on the definition of post dialing delay. Call set-up time is perceived by the user as the responsiveness of the service. Other factors such as ease of use contribute also to the user experience. The first of these factors is objective, the second is subjective [16].

Within the broad category of call quality two major factors contribute to the overall QoS experience of the user of the videoconferencing system:
- the end-to-end delay which impacts the interactivity of a conversation and
- the end-to-end video and speech quality.

The following factors contribute to the overall call-set up time:
- IP access network set up delays (these would include transport layer set up-times, modem training times and log on times at the ISP Gateway),
- signaling delays across the IP backbone,
- call set-up delays within the gatekeeper(s),
- access times and call processing delays to back-end services, such as directory services or authentication services,
- call set-up delays within the gateway,
- call set up times in the network(s).

The end-to-end delays are influenced by IP terminal buffering delays, H.323 packetization/buffering delays, codec delays and network transmission/propagation delays. The end-to-end audio/video quality depends on input and output devices, analogue/digital and digital/analogue circuit noise, video and audio coding distortion, effects of bandwidth limitation in the IP network.

The QoS issues associated with the IP terminal are the choice of codecs used in the terminal, the performance of the codec to various types of network degradation, the signal processing delays, the call processing delays, the number of frames per packet, the processing delays associated with security issues, the design of jitter buffers, the delays through the audio or digital media paths and the performance of echo-canceling devices [7], [8], [11].

1.2 QoS Issues Associated with LAN Access

In this configuration the access layer is limited to the Network Interface Card (NIC) used within the IP terminal. Though the LAN has ample bandwidth for transmission of coded speech/video, a fundamental issue frequently encountered is contention for shared media.
At any time, other (non audio) endpoints on the LAN, may flood the LAN and consume all the available bandwidth. This problem can only be avoided if there are mechanisms to manage and police the use of bandwidth (both for real-time use and best-effort use). The Subnet Bandwidth Manager (SBM) and RSVP (IETF RFC 2205) are intended to provide this capability.
The factors affecting QoS in this scenario are the transmission delays through NIC and the jitter in data buffers associated with the NIC. It is anticipated that these parameters will in general be well controlled and specification of upper bounds on these parameters should present few difficulties.

1.3 QoS Associated with PSTN Modem Accesses

In this type of access, modems are used to establish a digital channel between the videoconferencing terminal and the IP network. The factors affecting QoS in this configuration are the:
- modem bit rate,
- modem transmission overheads,

- throughput delay in modem and at ISP site,
- jitter within client modem, ISP modem and buffers,
- PSTN set-up time,
- modem connection set-up time,
- ISP logon & set-up time and
- error rate on PSTN link.

1.4 QoS Associated with ISDN Accesses

ISDN access uses a set bandwidth for the communication channel (16 kbit/s for the D channel, 64 kbit/s for a B channel). Aggregation of n*B channels to provide a 384 kbit/s channel provides a means of using video codecs even with normal RTP/UDP/TCP/IP overheads. The factors affecting QoS in this scenario are the:
- use of PPP/IP/UDP/RTP header compression on access link,
- throughput delay in ISDN terminal adapter and at ISP site,
- jitter within ISDN terminal adapter and ISP network interface buffers,
- ISDN set-up time and
- ISP Logon and session set-up time.

1.5 QoS Issues Associated with xDSL Accesses

xDSL access allows the use of various sizes of bandwidth, up to tens of Mbit/s, depending on application and the DSL technique used (e.g. ADSL, VDSL). IP access may use in general a mediation transport layer (i.e. ATM) or may be mapped directly into the xDSL frame (not standardized yet). The factors affecting QoS in this scenario are the:
- xDSL modem available bit rate (due to line condition and specific application),
- use of PPP/IP/UDP/RTP Header Compression on access link,
- throughput delay in xDSL modem (fast or interleaved) at ISP site,
- jitter within client modem, ISP modem and adaptation buffers,
- xDSL set-up time (e.g. when using Dynamic Power Save in VDSL application),
- ISP Logon and session set up time and
- error rate on access link.

2 Typical QoS Measurement Campaigns

2.1 Intrusive Measurement Tool = QoS Simulation Platform

The main functions of the requested tool are :
- capture and analysis of the received traffic at ATM, IP and application level, taking into account the different network architectures (IPoATM, LANE/MPOA, MPLS, etc.) and their characteristics,
- performance measurement at different network layers, focusing on relevant metrics such as: throughput, one-way delay, delay variation, packet loss, etc as referred in the RFCs 1944 and 2330,

- objective QoS evaluation, by taking into account the relevant performance parameters and network characteristics (e.g. scheduling techniques like WFQ or CBQ),
- generation of calibrated multimedia traffic patterns made of RSVP messages,
- traffic allocation to QoS Service Classes and CoS contract verification of IP streams based on Tspec parameters using a token bucket algorithm as referred in the RFC 2215. This verification would allow these streams to fit a DiffServ or IntServ network requirements,
- comparison with subjective data relating to the quality as seen by the end-users.

Iterative experiments will allow the tool to increase its knowledge in mapping objective and subjective QoS, to be able to allocate automatically the received traffic to the corresponding Service Class, only by measuring the adequate network performance. The simulation tool will enhance its QoS evaluation process by implementing innovative self-learning techniques.

2.2 Non- intrusive Measurement Tool = QoS Monitoring System

The monitoring tool (QoS Probes + Supervision System) will have following functions :
- processing the self-learning algorithms preliminary established by using the Simulation Tool in similar network environment,
- accessing the network in real-time, without disturbing the current traffic,
- traffic allocation to QoS Service Classes,
- results supervision via data collection by a supervision system, which will be designed to be connected, via TMN interfaces, to the network administration system,
- supervision system design taking into account entities from DiffServ architecture like Bandwidth Brokers (BB) or protocols like Common Open Policy Service (COPS).

2.3 End-to-End QoS Experiments

Experimental services will be based on multimedia applications. Voice over IP quality will be determined according to G.711 64kB/S characteristics. Video over IP will refer to H.320 ISDN video-conferencing running at 384KB/S. QoS expertise will take into account most types of coding and compression algorithms (G.723, H.323, M/J-PEG.).
Performance will be measured at standardized access points :
- PCM n*64KB/S,
- ISDN, basic and primary accesses,
- PDH, SDH (STM1-STM4),
- xDSL subscriber loop accesses (probably ADSL),
- Cable-modem accesses,
- ATM (PDH 34MB/S, SDH STM1-STM4).

The evaluation process will make use of self-learning techniques, allowing statistical approach of QoS parameters. QoS expertise will take into account real-time enhancements of transport protocols (UDP/IP, RSVP, RTP, RTCP, IPV6...). As QoS management involves all terminal and network related aspects, measurement and monitoring will cover most of them. These include Tspec definition, Adspec definition if RSVP is to be used, RSVP/ATM issues, DiffServ/IntServ interface issues, DiffServ/ATM issues, MPOA/LANE QoS if used, IPv6/ATM, etc... Measurement techniques will be enhanced in order to anticipate performance degradation and to predict QoS evolution [2], [3].

Performance evaluation will be characterized by following aspects:
- connections opening and closing mechanisms,
- data transfer mechanisms,
- addressing, parameters negotiation,
- congestion and errors control,
- bandwidth allocation,
- CoS contract verification.

3 Voice over IP Experiments

3.1 ETSI TIPHON Project Simulation Experiments

The ETSI TIPHON contribution 09TD42 presented a proposal to carry out simulations where different scenarios of end to end speech transmission over IP based networks could be evaluated with respect to speech transmission quality.
In autumn 1998, an investigation according to 09TD42 was performed by Deutsche Telekom Berkom (T-Berkom) where such different scenarios were simulated and subjectively assessed in a well-established listening. The simulation processing contained a couple of speech codecs, packet loss ratios and various kinds of audio frames per IP packet. This TD describes the test methodology, the simulation method, the scenarios and the results of the executed simulation processing.
Furthermore we would like to discuss some interesting results concerning the relationship of speech material (construction, length) packet loss and their influence to the auditory assessment.

3.1.1 Test Methodology

TIPHON WG5 (DTR/TIPHON 05001 V1.2.5, chapter 7.3.2) defined a methodology for testing speech quality in TIPHON compliant networks and terminals. This methodology was taken into account and used as a basis model for the T Berkom simulation processing.

A set of speech signals designed according to ITU-T Rec. P.800 was used as input of the simulation path. The simulation path includes the terminal side (electrical part) and the network itself. The influence of the terminal side was focussed to the speech conversion and IP packet size issue. The influence of the network side was simulated by different packet loss rates. After the simulation the speech samples were recorded and stored in a database.

3.1.2 Simulation Method

For simulation of network influences in the case of packet loss, a common channel model was designed, realized by channel files which describe the network condition with the same time resolution as the source speech sample rate. So the network has a certain condition (good or bad) for every speech sample (every 125 µs) two adjacent network states were considered as statistically independent, because the network speed was assumed to be much higher than the sample rate (8000 samples per second). So for each packet loss rate one channel file was created using a random generator. The length of this channel file was exactly the same as the length of the speech file.

In a further step the speech file, assembled in IP packets, was matched to the channel file. According to the length of the IP packet (10ms, 20ms,...) the channel file was checked every time when a packet was ready to send. That means if the packet size was 10 ms the channel file was checked also every 10 ms if the condition is good or bad. In a bad case the IP packet was lost, otherwise it was further processed. This information (IP packet lost or not) was stored in a description file which was the input of the re-assembler and speech decoder.

3.2 Testing of Speech Quality

There are two methods of testing end-to-end (acoustic to acoustic) speech quality:
- subjective tests involving the opinion of panels of users (see ITU-T Recommendation P.800),
- objective tests including comparison methods against a known reference signal (See ITU-T Recommendation P.861), absolute estimation methods.

Based on ITU-T Recommendation P.561, the measurement of individual parameters followed by the use of a Transmission Rating Model (TRM) to combine the effects of the individual parameters and predict the subjective views of users. The E-model is under consideration for this purpose.

Subjective tests have the advantage of including all parameters and providing a direct subjective view, but they take a long time to perform, are costly and are ill-suited to investigating changes in the values of many parameters because of the large numbers of combinations involved.

Objective comparison methods are described in DEG/STQ00001. Objective tests using the E-Model approach should include the same parameters as in the PSTN world:

- SLR   Sending Loudness Rating;
- RLR   Receiving Loudness Rating;
- OLR   Overall Loudness Rating;
- STMR   Sidetone Masking Rating;
- LSTR   Listener Sidetone Rating;
- Ds   D-Value of Telephone at Send-side;
- Dr   D-Value of Telephone at Receive-side;
- WEPL   Weighted Echo Path Loss;
- qdu   Number of Quantizing Distortion Units;
- Ie   Equipment Impairment Factor (low bit-rate Codecs);
- Nc   Circuit Noise referred to the 0 dBr-point;
- Nfor   Noise Floor at the Receive-side;
- Ps   Room Noise at the Send-side;
- Pr   Room Noise at the Receive-side.

For evaluation of the Ie values for low bit-rate codecs, some objective measurement methods have been developed but commercial measurement systems are not yet available. In addition, specific requirements from the TIPHON system (eg. packet loss) have to be considered in determining Ie.

In conversational situations:
- TELR   Talker Echo Loudness Rating;
- T   Mean one way delay of the echo path; and
- Tr   Roundtrip Delay in a closed 4-wire loop,
need also to be considered.
The performance of TIPHON systems in terms of TIPHON speech quality classes may also be measured between the electrical input/outputs of the TIPHON terminals or SCN telephone terminals connected to the TIPHON system. Figure 2 shows in general how this should be done.

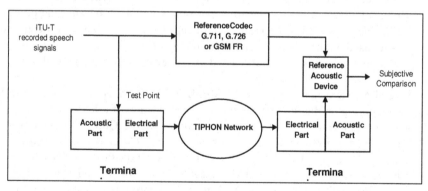

Fig. 2: Methodology for testing TIPHON speech quality

Speech quality shall be measured using the subjective test methodology as defined by ITU-T SG12 until such times as calibrated objective methods are possible. It is planned that these test results will be used in the future to enable predictions of overall performance to be made using a TRM (e.g. the E-Model). It should be noted that the E-model is not a test method.

4 Desktop Videoconferencing Experiments

4.1 State of the Art

4.1.1 State of the Art of QoS in IP Infrastructures

For many years, public network operators regarded ATM as *the* solution for a service integrating broadband network. Conceived as a logical extension of narrowband ISDN, its standardization was influenced by the connection-oriented paradigm, signaling protocols and addressing scheme known from ISDN. While ATM research, development and standardization was concerned with guaranteed QoS for typical broadband ISDN applications, such as video-on-demand and multimedia conferencing as well as with an efficient rate control for data applications (Available Bit Rate), the World Wide Web helped to establish IP networks as *the* carrier for data networking.

The current Internet architecture offers a flexible, but simple connectionless best effort service and this is inadequate for applications sensitive to the QoS provided by the network. For this reason, the IETF has been working on extensions to the current Internet protocol suite in order to enable service guarantees (Resource Reservation Protocol RSVP together with Integrated Services, IIS) or at least differentiation (Differentiated Services, DS).

In Integrated Services, RSVP allows applications to request either Guaranteed or Controlled Load Service for individual flows in an IP network. However, the wide scale deployment of RSVP must be approached with care because the processing of (periodically refreshed) reservation and control messages, the identification of each packet based on the IP header and the handling of per-flow reservation state becomes challenging in backbone routers passed by a huge number of individual flows.

Several ACTS projects (DIANA, SUSIE, BTI, PETERPAN, IthACI and ELISA) have been working on the implementation and optimization of a RSVP over ATM control architecture to integrate IP and ATM while enforcing QoS end-to-end. Their work extends standardized solutions to integrate IP and ATM on a best-effort basis, namely Classical IP over ATM (CLIP), Next Hop Resolution Protocol (NHRP), Multicast Address Resolution Server (MARS), LAN Emulation (LANE) and Multiprotocol over ATM (MPOA) that focus on IP to ATM address resolution to set-up switched best-effort ATM VCs.

The resulting RSVP over ATM architecture is an example of traffic descriptor and QoS parameter based resource reservation that guarantees tight QoS end-to-end. The aforementioned scalability issues are addressed by applying a concept of massive aggregation of flows to a single VC.

In contrast, Differentiated Services architecture achieve scalability by classifying and marking packets by means of the so-called DS field in the IP header once at the ingress to a DS capable IP network. Based on a (relatively static) Service Level Agreement profile negotiated between an Internet provider and an user, traffic will receive a particular per-hop forwarding behavior on routers that interpret the DS field. Without explicit signaling and admission control, DS have the potential to provide relative QoS to so-called behavior aggregates (streams that are marked with the same priority) using simple and scalable mechanisms.

Hence, the Internet is evolving in the direction of a multi-service network that supports several traffic classes, signaling and various other attributes associated with a stream of data. Since various solution will co-exist, QoS applications and services have to build upon a more generic protocol layer, as represented by the H.323 series of recommendations, that is to be translated to native network layers and traffic control capabilities in terminals (with operating systems support) or gateways (network equipment vendors). This approach enables application developers to keep pace with and to make use of the rapidly evolving IP based technologies and network infrastructures.

4.1.2 State of the Art in QoS Evaluation Methodologies, Tools, and Experiments

Today, many of activities address QoS evaluation methods and tools dedicated to broadband networks ; most of them concentrate on objective (quantifiable, measurable) QoS aspects, which are mainly network oriented. Subjective aspects relate to the user's point of view and are only approached qualitatively.

Objective QoS

The IETF's work on Quality of Service and performance is primary directed towards developing new protocols which will allow a degree of bandwidth reservation or Quality of Service differentiation. The IP Performance Metrics working group (IPPM) is working on defining a set of standard metrics that can be used to derive a quantitative measure of the quality, performance and reliability of Internet data delivery services.

Measurement and monitoring processes will also take into account work performed in :
- ACTS Project ISABEL: video-conferencing traffic characteristics, ISABEL and Mbone tools,
- Project MEHARI (Spanish local project): techniques and tools for the analysis of Internet services. Currently the MEHARI system is used to do measurements on IP over ATM. New functionality could be as for example LANE/MPOA measurements, end-to-end QoS monitoring, etc.

- Project SABA (Spanish local project): new services and protocols for the Broadband Spanish Academic Network. It is a project on next generation Internet (QoS management, terminal and network related aspects).

Subjective QoS

QoS requirements by the user/customer is the statement of the level of quality of a particular service required or preferred by the user/customer. The level of quality may be expressed by the user/customer in technical or non-technical language.

A typical user/customer is not concerned with how a particular service is provided or with any of the aspects of the network's internal design, but only with the resulting end-to-end service quality. From the user's/customer's point of view, QoS is expressed by parameters that:

- focus on user/customer-perceivable effects, rather than their causes within the network,
- do not depend in their definition on assumptions about the internal design of the network,
- take into account all aspects of the service from the user's/customer's point of view,
- may be assured to a user/customer by the service provider(s),
- are described in network independent terms and create a common language understandable by both the user/customer and the service provider.

Classes of Service

The terminology Class of Service is used to describe a scheme where service types are grouped logically together. This grouping can then be used as the basis for service type prioritization.

Four classes of service are defined by ETSI:
- Class4 : Best quality,
- Class3 : High Quality,
- Class2 : Medium Quality,
- Class1 : Best Effort Quality.

Different (and complementary) approaches are being defined in the two main QoS architectures of the IETF [5]. These architectures will be prevalent in the near future Internet:
- Differentiated Services (DiffServ):
  . Default (DE): best-effort,
  . Assured forwarding (AF): not completed yet,
  . Expedited forwarding (EF): not completed yet.
- Integrated Services (IntServ):
  . Guaranteed: with bandwidth, bounded delay and no-loss guarantees,
  . Controlled load: simulates a best-effort service in a lightly loaded network,
  . Best-effort.

One important issue under work is the mapping of CoS definitions of both standards IntServ and DiffServ.

4.1.3 State of the Art of Multimedia Conferencing Systems

Almost all multimedia or video conferencing systems used today are either based on IP (e.g. CUSeeMe, the MBone tools and Microsoft's Netmeeting) or on ISDN (e.g. Intel ProShare and PictureTel). This leads to a conservative dimensioning where the application on the one hand must be able to deal with data losses and on the other hand must put restrictions on its output bit rate to protect the other services running over the same network from excessive losses.

This scheme works very well for data, but is problematic for real-time audiovisual services. Furthermore the available bit-rate on the Internet is quite low which leads to designs with low resolution (image size) and low frame-rates. With the use of ATM these problems could be reduced or eliminated.

Multimedia conferencing systems using native ATM are not commercial available today, but some research projects have worked in this area. The RACE project R2025 MIMIS have worked on multimedia desktop conferencing for ATM networks, using the early drafts of the ITU-T T.120 series.

4.2 The DIVINE Project

DIVINE is an European consortium regrouping industrial companies, research centers, telecommunications operators, universities and end-users [4]. The DIVINE project was initiated by the European Commission (DGXIII) in the frame of the 4th PC&RD ACTS Program.

The main objectives of the DIVINE project were:
- to be a "market-driven" project through operational field trials involving real end users, to issue meaningful conclusions on the viability of the deployment of multi-point multimedia applications on high speed communication infrastructures,
- to be a "product-oriented" project : the reuse of the results of the project is a key issue. It implies that the DIVINE system has to fit with the user requirements related with functionality, quality of service and price,
- to demonstrate the interoperability between the B-ISDN DIVINE and N-ISDN commercial videoconferencing products,
- to demonstrate on a large scale experimentation the interoperability of European ATM-LAN with the ATM-WAN,
- to promote the standards in broadband desktop videoconferencing applications and to contribute to the standardization work in IMTC, ETSI or ITU-TS [6].

Bandwidth-on-demand should be easily and flexibly controlled by the user. Support of a variety of interface-cards for ATM, graphics, video-input and video compression should be integrated. Access to these peripheral components should be enabled via hardware interfaces. If such an interface does not exist, at least a standardization should be considered or this should be open to the public.

The DIVINE protocol stacks can be described as follows:

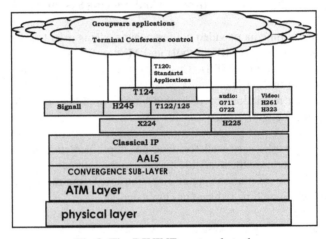

Fig 3: The DIVINE protocol stack

Unicast and multicast videoconferencing is requested, as well as easy interworking with ISDN based videoconferencing facilities.

4.3 Tests and Performance Measurements

The host configuration includes machines, graphic interfaces, audio/video codecs and the network access interfaces includes LAN Ethernet: 10 Mbit/s, 100 Mbit/s, Fast Ethernet, Switched Ethernet, 155 Mbit/s ATM. Quality of Service evaluation relates to end-to-end multimedia applications with LAN-to-LAN interconnection, audio/video conferencing and audio/video transmission and distribution.

The main test functions are the user/network protocol monitoring, the user/network protocol simulation, the traffic load simulation and the errors insertion (bit/cell/frame errors, delays, jitter...).

In user simulation mode, the ATM analyzer acts as any number of user devices which may communicate with a real ATM switch under test. A real ATM switch is attached to the analyzer which may simulate any number of user devices which are called virtual stations. Additionally, the analyzer enables the simulation of any number of user devices which may communicate with the simulated switch or with any other user device. The measurements include:
- estimation of bandwidth allocated to a videoconferencing session,
- videoconferencing behavior under strong network load conditions,
- audio/video quality estimation,
- determination of optimal Quality/Bitrate relations,

- interoperability with ISDN (H320/H323 gateway tests),
- native ATM interworking (ATM access for DIVINE terminal).

Objective measurements were done essentially by observing traffic at the ATM level (Radcom monitoring equipment) and at the Ethernet level (Meterware monitoring equipment), via cable-modems and via ISDN H320 gateway [15].

During our tests, both videoconferencing applications are adjusted for maximum video quality, (the maximum bandwidth allocated for DIVINE is 750 kbit/s). Statistics are extracted with the RADCOM test equipment, connected to the ATM network and the results are depicted for a 1 minute connection (data storage capabilities).

The results can be represented in this following figure:

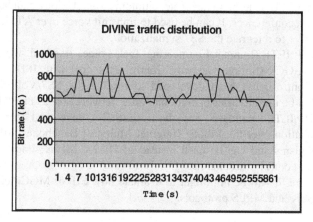

Fig. 4 : DIVINE traffic distribution

Decreasing available bandwidth affects video and audio quality, but the connections are not lost. The maximum quality requests only 450 kbit/s to 900 kbit/s bandwidth allocation and videoconferencing applications request a minimum of available bandwidth to provide acceptable video and audio quality.

Fugitive traffic congestion disturbs the video transmission for short periods of time, but does not affect the videoconferencing session itself. When the available bandwidth increases again, video and audio quality recover their initial values.

5   The Future of VoIP Networks

New operators use more and more VoIP because they can use the same equipment to transmit voice and data over Internet. The development of VoIP implies the integration of the PSTN (Public Switched Telephone Network) network. Therefore gateways that can be used to interconnect SS7 protocol with IP protocol are now available. The three majors protocols for VoIP and SS7 over IP are H.323, MGCP (Media Gateway Control Protocol) and SIP (Session Initiative Protocol).

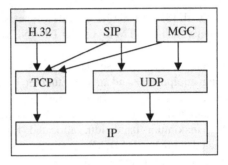

Fig. 5: VoIP and SS7 over IP protocols

The H.323 ITU (International Telecommunication Union) standard is adapted for multimedia conferences. It can be used to transmit voice over ATM, but H.323 cannot really evolve to integrate the SS7 signalization.
The MGCP IETF (Internet Engineering Task Force) standard, which come from the merge of SGCP (Simple Gateway Control Protocol) with IPDC (Internet Protocol Device Control), has been developed to resolve the SS7/VoIP integration. Then, MGCP can be used to offer operational VoIP networks based on PSTN over IP.
The SIP IETF standard has been initially developed for the multimedia communication over the Mbone (Internet Multicast Backbone). SIP offers SS7 over IP mechanisms and can be used instead of H.323. SIP is simpler than MGCP that offer more control mechanisms.
In the future, it will be important to evaluate how SIP or MGCP can be managed by the DiffServ and MPLS protocols.

6 Conclusion

The QoS for VoIP systems and desktop videoconferencing is more and more enhanced. In this article, we present some methods and performance measurements applied to VoIP and to desktop videoconferencing systems. Some methods for the simulations and for the tests are experimented in different projects.

7 References

1. K. Achtmann, K.H. Doring, R. Herber, G. Komp, "An ATM-based demonstration model for multimedia services using different access networks", Multimedia Applications, Services and Technologies, ECMAST'97, May 21-23, 1997, Milan, Italy, 1-17.
2. G. Armitage, "MPLS: the Magic behind the Myths", IEEE Communications Magazine, Vol. 38, No 1, , January 2000, 124-131.
3. R. Cocca, S. Salsano, M. Listanti, "Internet Integrated Service over ATM: a Solution for Shortcut QoS Virtual Channels, IEEE Communications Magazine, Vol. 37, No 12, , December 1999, 98-104.

4. DIVINE (Deployment of Interpersonal Videoconferencing Systems on IBC Networks) project, ACTS European Project AC035, 1998.
5. G. Eichler, H. Hussmann, G. Mamais, I. Venieris, C. Prehofer, S. Salsano, "Implementing Integrated and Differentiated Services for the Internet with ATM Networks: a practical approach", IEEE Communications Magazine, Vol. 38, No 1, January 2000, 132-141.
6. ETSI/DTR/TIPHON-05001 V1.2.5., "Telecommunications and Internet Protocol Harmonization Over Networks ; general aspects of Quality of Service", 1998
7. B. Goodman, "Internet Telephony and Modem Delay", IEEE Network, Vol. 13, May-June 1999, 8-16.
8. A.M. Grilo, P.M. Carvalho, L.M. Medeiros, M.S. Nunes, "VTOA/VoIP/ISDN Telephony Gateway", 2nd International Conference on ATM, ICATM'99, Colmar, June 21-23 1999, 203-235.
9. ISO/IEC DIS13236 , "Information Technology Quality of Service - Framework," 1996.
10. N. Kroth, L. Mark, J. Tiemann, " A Framework for Testing IP QoS over ATM Networks: Implementation and Practical Experiences", 2nd International Conference on ATM, ICATM'99, Colmar, June 21-23 1999, 203-235.
11. V. Mirchandani, D. Everitt, "Performance of integrated telephony service over standalone LAN and ATM interworked LANs", 22nd Conference on Local Computer Networks, Minneapolis, USA, November, 2-5, 1997, 80-88.
12. D. Newman, "VoIP Gateway: voicing doubts", Data Communications International, Vol. 38, No 12, September 1999, 70-78.
13. Z. Peifang, "Scalability and QoS guarantee in IP networks", 8th International Conference on Computer Communications and Networks, 627-633.
14. X. Scharff, P. Lorenz, "Specification of a Multimedia Application Generator in Telecommunication Systems", 12th International Conference on Computer Applications in Industry and Engineering (CAINE'99), Atlanta, USA, November 4-6 1999, 54-57.
15. H. Tobiet, P. Lorenz, "Performance measurements on an ATM-based Metropolitan Area Network: OASICE Case Study", 1st IEEE International Conference on ATM, ICATM'98, Colmar, June 22-24 1998, France, 410-417.
16. J. Toga, J. Oltt, "ITU-T Standardization Activities for Interactive Multimedia Communications on Packet-based Networks: H.323 and related recommendations", Computer Networks, Vol. 31, No 3, February 199,9 205-223.
17. M. Valino, J. Corchado, "VoIP: the convergence of Networks", Computing and Imformation Systems, Vol. 6, No 3, October 1999, 105-112.
18. P.P. White, "ATM switching and IP Routing Integration: the Next Stage in Internet Evolution", IEEE Communication Magazine, April 1998, 78-83.

Efficient Network Utilization for Multimedia Wireless Networks

Xin Liu, Edwin K.P. Chong, and Ness B. Shroff

School of Electrical and Computer Engineering
Purdue University, Lafayette IN 47907, USA
Tel: +1 765 494-1744, Fax: +1 765 494-3358
{xinliu,echong,shroff}@ecn.purdue.edu

Abstract. In this paper, we present an access scheme to satisfy the QoS requirements for two classes of traffic during the *contention phase* of packet-switched wireless communications. In the proposed scheme, different classes of users contend with other users for resources based on controlled class-dependent permission probabilities. We prove that our algorithm is stable for a large class of arrival processes. Under certain QoS requirements, we derive an upper-bound on the throughput for a general class of random access algorithms. We show that the throughput of our algorithm asymptotically approaches this upper-bound. We also show, through numerical examples, that our algorithm achieves high network utilization.

1 Introduction

The goal of wireless communications is to provide a convenient and economical way for people to transfer all kinds of information, such as voice and data. Compared with circuit switching, packet switching provides more efficient multiplexing of different classes of traffic. In circuit switched networks, when a user is admitted to the network, a certain amount of network resource is assigned to the user and exclusively used by the user until its communication finishes, regardless of whether the user has information to transmit during this period. In packet switched networks, when a new user is admitted, no specific resource is assigned to it. Resources are shared by users in the system. A user only occupies the network resource when it has information to transmit. Consider a phone call as an example. When the user talks, voice packets are generated at a certain rate; when the user is silent, no voice packet is generated. On average, the user talks less than half of the entire call duration. In circuit switched networks, the networks assign the voice user the resource equivalent to its packet rate during talking, hence about half of the resources are wasted. In packet switched networks, when a user does not talk, no resource is assigned to this user; when the user begins talking after a period of silence, the network assigns resource to this user again. Hence, in general, packet switching utilizes network resources more efficiently than circuit switching. Efficiency is very important for wireless networks because wireless bandwidth is scarce. However, wireless packet switching

C.G. Omidyar (Ed.): MWCN 2000, LNCS 1818, pp. 108–122, 2000.

suffers from access problems in the uplink. In other words, when a user becomes active, it has packets to transmit and no network resource is assigned to it, the user has to compete with other users to gain the access to network resources. To solve this problem, a variety of contention and reservation medium access control (MAC) protocols have been widely used in the area of communication networks [2, 3, 4, 5]. Typically, there are two transmission phases:

1. Newly activated users compete to gain access to the networks. The first packet of a newly activated user is transmitted through the network using some random access protocols; i.e., contention-based communications. This first packet may be a packet in a special form or a normal data packet. In this paper, we call the first packet a *request*. If the first packet is lost during transmission, or is received in error, then it is retransmitted until successful.
2. Following the first successful contention-based transmission, subsequent transmissions are scheduled contention-free using a scheduling strategy.

We call the first phase the *contention phase* and the second phase the scheduling phase. In this paper, we focus on the contention phase of communications. In packet switched wireless networks, the contention phase may exist throughout the whole communication period, and not only during the admission period. Every time a user becomes active (say, a user begins talking after being silent), at that very moment, because no resource is assigned to the user, the user has to inform the base station about its resource requirement through contention-based communication. Hence, contention-based communication plays an important role in packet-switched wireless networks.

In packet switched networks, admission control and resource allocation are used to provide QoS. In general, admission control is based on the resource allocation scheme. In wired networks, resource allocation is implemented by smart scheduling schemes. However, smart scheduling is not enough to provide QoS for wireless networks, where contention plays an important part. For example, we want to provide delay guarantee to real-time traffic in wireless networks. When a user begins talking, it first sends its request to the base station through random access; i.e., contention-based transmission. Then the base station schedules the traffic after it receives a resource request from the user. Therefore, the user experiences delay caused by contention plus the delay caused by scheduling. To guarantee the delay experienced by the user, we need to guarantee the delay in both contention phase and scheduling phase. During the scheduling phase (if one actually exists in the given implementation), smart scheduling strategies can be used to provide delay guarantees. However, we also need algorithms in the contention phase to provide delay guarantees to users. To provide QoS in the contention phase is intrinsically difficult due to the nature of random access. While there is a significant body of work on the development of effective scheduling and admission control policies to ensure QoS, there is very little work done in implementing QoS during the contention phase of communication.

In this paper, we present an algorithm that implements QoS requirements for two classes of traffic in the contention phase of packet switched time-slotted wireless networks. Controlled time-slotted ALOHA is the random access algorithm considered in this paper. Two traffic classes, voice and data, are considered. We consider only two classes for simplicity of exposition, convenience of calculation and explanations, although more classes can be similarly considered. We assume that voice users have *delay requirements* and data users do not have such requirements.

In wire-line networks, if two or more users transmit at the same time through the same media, usually all of the transmissions are assumed to have failed. However, this assumption may be unnecessarily pessimistic in the mobile radio environment, where the received packets at the base station are subject to the near/far effect and channel fading. Packets from different users in the same slot may arrive at the base station with different power levels and the base station may successfully decode one or more packet. This is referred to as *capture*. Due to the page-limit requirement of this conference, we only present the QoS algorithm for systems that do not exploit capture. However, the proposed algorithm works for systems with capture too [6]. It is obvious that the system throughput will be improved if the system explores capture. However, unfairness exists between near and far users due to the nature of radio transmission. To achieve fairness and good throughput, we present a distance-dependent permission probability scheme that require users at different distances from the base station to transmit with different probabilities to provide certain delay guarantees, *distance fairness*, and good throughput. In summary, if we do not consider the ability of capture, the QoS requirement is presented in terms of delay. When we consider capture, the QoS requirement is explained in terms of delay and distance fairness.

This paper is organized as follows. In Section 2, we describe the system model. We present and analyze the QoS algorithm in Section 3. An upper-bound for the throughput is derived, under certain QoS requirements, for a general class of random access algorithms. The throughput of our algorithm asymptotically approaches this upper-bound. Simulation results are provided in Section 4. Conclusion and future work are presented in Section 5.

2   System Model

In this section we describe the system model. There is a base station with mobile users in its coverage area. We consider the uplink of a time-slotted system and focus on the contention phase of communication. We assume that time is divided into frames and each frame consists of M request slots. Each request slot is large enough to contain a fixed size request. The base station monitors and controls the contention phase in the system. *In the following, when we mention users we mean newly activated users with requests to transmit, except otherwise specified.*

At the beginning of a frame, the base station broadcasts a permission probability for each class of users through a non-collision error-free signaling channel. A user decides whether or not to transmit in a request slot in the frame according to the permission probability of its class broadcasted by the base station. Different classes of users may have different permission probabilities.

We assume that a user can transmit at most once in a frame. There are M request slots in each frame. The parameter, M, determines how often the base station updates its control parameters, and how long a user waits before it retransmits. In practice, the larger the value of M, the less the signaling, the better the estimation of the number of users, however, the longer the delay.

In some cases, we prefer a large value of M. An example of such a scenario is in satellite communications. After the contention of a time slot, a user cannot know immediately whether its request is successfully received by the hub station. In satellite communications, the round trip delay is relatively large. For instance, the propagation delay is around 20–25ms for LEO (low earth orbit) systems [8]. An immediate ack from the the hub is impossible. Furthermore, the coverage area of satellite communications is relatively large, it is difficult for an earth station to detect whether its transmission is successful. Hence, a large value of M may be suitable for such a case. In other cases, a small value of M could be favored. A good example of such a case is a local wireless network, where the sum of the round trip delay, and processing time, etc., is small. A user transmits, then waits for acknowledgment. If the user does not receive an acknowledgment from the base station in the predetermined waiting time, it assumes that the transmission has failed. The user could retransmit it in the next frame. The extreme case is when $M = 1$; i.e., a user can retransmit its request in the next request slot. In the extreme case $M = 1$, the scheme studied in this paper becomes the pure priority scheme; i.e., when there are voice users, no data user transmits, and when there is no voice user, data users transmit. However, even in a wireless LAN, it is not necessary to adopt such a small value of M. Usually, the requests are much shorter than normal data packets. Hence, the delay caused by several request slots are tolerable in order to reduce the cost of extensive signaling.

In this paper, we assume that the system is not capable of correctly deciphering any transmissions when two or more overlapping transmissions arrive in the same slot; i.e., if two or more users transmit their requests through the same request slot in a frame, neither of them can be successfully received. This situation is called collision.

We assume that a request is never discarded; i.e., a user always retransmits its request until it is acknowledged by the base station that its request has been received successfully. While the request of a user is delayed, some packets may be buffered at the user. In real-time applications, human factors may decide whether to send a delayed packet or to drop it. This issue is irrelevant in our scheme. Furthermore, we assume that the acknowledgment is error-free and the base station uses a scheduling strategy to decide when the active user should transmit in the reservation phase of communication.

3 The QoS Algorithm

We first present the QoS algorithm with restriction to the delay requirement of voice users. We, then, analyze the throughput and stable condition. Finally, we derive a throughput upper-bound under the QoS requirement for a large class of random access algorithms.

3.1   Algorithm

Let p_v (p_d) denote the permission probability that a voice (data) user transmits in a request slot in a frame. In this paper, the permission probabilities, p_v and p_d, are used to stabilize the ALOHA system, to achieve good throughput, and to provide QoS guarantees. The use of permission probabilities to stabilize ALOHA is not a new idea. Permission probabilities are also used to provide priority to voice users in [4, 7]. In the literature, there are algorithms, centralized and decentralized, to estimate the number of users in the system. All these algorithms can be used in our scheme. Hence, we focus on how to use the permission probabilities to satisfy QoS instead of how to estimate the number of users. During the analysis we assume that the base station knows the precise numbers of voice users and data users in each frame. Knowing this information is the ideal condition of the algorithm. Practically, we use a Kalman filter to estimate the numbers of voice users and data users with requests in each frame. We show through simulations that using a Kalman filter for the estimation provides very good results.

As mentioned before, a user can transmit at most once in a frame. We do not distinguish between newly arrived and retransmitted users. The base station broadcasts p_v and p_d at the beginning of frame i. A voice user randomly selects a request slot to transmit in this frame with probability p_v, as would a data user with probability p_d. All users select and transmit independently. The base station acknowledges those users whose requests have been successfully accepted at the end of frame i. Users that have not been acknowledged assume that their requests have not been successfully transmitted. They retransmit in the next frame. The base station estimates the number of users in the system, calculates p_v and p_d for frame $i + 1$, and so on. It is easy to prove that the throughput is maximized when M users transmit in each frame [6]. However, this throughput may come at the cost of excessive delay for voice users. Hence, we need to develop a scheme that attempts to maximize throughput subject to a given level of delay requirement for voice users.

A good measure of QoS is the delay experienced by a user before its request is successfully received by the base station. However, the precise delay distribution of voice users is very difficult to find in this context. Thus, we define an average success probability, $\bar{P}_s$, as the QoS measure used in this paper. Suppose the system has reached steady state. When a voice user becomes active, on average, it transmits its request successfully with probability $\bar{P}_s$, given by

$$\bar{P}_s := E\left[p_s(N_v, N_d)\right] = \sum_{i,j} p_s(i, j)\pi(i, j), \tag{1}$$

where $p_s(i, j)$ is the probability that a voice user transmits its request successfully in a frame in steady state when there are i voice users and j data users in the system, and $\pi(i, j)$ is the steady state distribution that i voice users and j data users are in the system.

Our QoS requirement for voice users is $\bar{P}_s \geq A_0$, where A_0 is the given delay threshold. Roughly speaking, the contention delay of a voice user is geometrically distributed with parameter $\bar{P}_s$; i.e., the distribution of access delay D is approximated by $P(D = x) = \bar{P}_s(1 - \bar{P}_s)^{x-1}$. When the correlation of the numbers of users between cells is small, the approximation is good. If the number of users arrived in each frame is independent, then the larger the M, the better the approximation. In Section 4, we show that the distribution of voice users from simulations is well approximated by a geometric distribution (see Figure 1).

The QoS algorithm is described as follows. Suppose that the base station knows that N_v voice users and N_d data users are in the system. Then, the permission probabilities of voice users and data users are

$$p_v = \min(1, \frac{M}{N_v}),$$

$$p_d = \begin{cases} \min\left(1, \frac{(C-N_v)^+}{N_d}\right) & : & \text{if } N_v > 0, \\ \min\left(1, \frac{M}{N_d}\right) & : & \text{if } N_v = 0, \end{cases} \qquad (2)$$

where

$$(x)^+ = \begin{cases} x & : & \text{if } x \geq 0, \\ 0 & : & \text{otherwise.} \end{cases}$$

Note that C is a tuning parameter used to satisfy the QoS requirements of voice users. So the algorithm does the following. If the number of voice users in the system is less than M, all voice users can transmit freely. In this case, data users may or may not be allowed to transmit. If the number of voice users in the system is greater than M, then a voice user is allowed to transmit based on the outcome of the toss of a biased coin with probability M/N_v of success. In this case, no data users are allowed to transmit. Before we illustrate how to calculate C, we first make a few observations:

- Data users yield to voice users the right to access request slots.
- The parameter C satisfies $0 \leq C \leq M$. The expected number of data users to transmit is $(C - N_v)^+$. The total throughput is maximized when $C = M$. The larger the value of C, the higher the throughput, and the larger the delay of voice users. Hence, there is a tradeoff between the throughput of the system and the delay requirement of voice users. When the QoS requirement is stringent, C is small, data users are allowed to access request slots with lower probability, and voice users have a higher probability to succeed in a frame.
- When there is no voice user; i.e., $N_v = 0$, the value of p_d is set to maximize the throughput.

The tuning parameter C can be calculated theoretically. A two dimensional Markov chain is used to calculate the steady-state distribution. Suppose that we know the distribution of the arrival process. Let $C = x, 0 \le x \le M$. Transmission probabilities between states are determined by (2) and the arrival process. Hence, $\pi(i, k)$ can be calculated and so can $\bar{P}_s$. Since $\bar{P}_s$ is a monotone decreasing function of x, denoted as $\bar{P}_s(x)$ for $0 \le x \le M$, the parameter C is the unique root of $\bar{P}_s(x) = A_0$, which can be obtained easily using standard zero-finding algorithms. If $\bar{P}_s(0) < A_0$, the QoS requirement cannot be satisfied. In other words, even without data users, the delay caused by the contention among voice users are still larger than required when $\bar{P}_s(0) < A_0$.

Practically, there is a very simple approximation for C. Let K_0 satisfy

$$\left(1 - \frac{1}{M}\right)^{K_0 - 1} = A_0. \tag{3}$$

If K_0 is not too small compared to M and the fraction of voice users is not too large, then K_0 is a good approximation of C. In this case, the number of voice users in the system in steady state is seldom larger than K_0. Therefore, the average delay $\bar{P}_s$ is:

$$\bar{P}_s = E(p_s) = E(p_s(i) | i \le C) p(i \le C) + E(p_s(i) | i > C) p(i > C)$$
$$\approx (1 - \frac{1}{M})^{C-1} = (1 - \frac{1}{M})^{K_0 - 1} = A_0.$$

In fact, if $K_0 \ge 0.5M$ and the fraction of voice users is less than 70%, $C \approx K_0$ is a good approximation. We set $C = K_0$ in simulations in Section 4 and find that it works well.

We, next, analyze the algorithm. First, we calculate the throughput. Second, we prove that the algorithm is stable for a large class of arrival processes. Then, we derive an upper bound on the throughput of random access algorithms under the QoS requirement $\bar{P}_s \ge A_0$. We show that the throughput of our algorithm asymptotically approaches the upper-bound.

3.2 Throughput

Suppose that there are k users transmitting in a frame. Each user selects one of the request slots randomly and independently. When only one user transmits in a request slot, we assume that the transmission is successful. When two or more users transmit in the same request slot, we assume that neither of the transmission is successful. The throughput, T_k, is defined as the average number of requests that are successfully transmitted in a frame and p_k is the probability that a user transmits successfully. We then have

$$T_k = k \left(1 - \frac{1}{M}\right)^{k-1},$$

$$p_k = \frac{T_k}{k} = \left(1 - \frac{1}{M}\right)^{k-1}.$$

We consider the throughput under three conditions:

1. When $N_v \geq C$, each voice user transmits in a request slot with probability $p_v = \min(1, M/N_v)$ and no data user transmits. The throughput is:

$$T(N_v, N_d) = \sum_{i=0}^{N_v} T_i P(i \text{ voice users transmit in this frame})$$

$$= N_v p_v \left(1 - \frac{p_v}{M}\right)^{N_v - 1}. \tag{4}$$

2. When $N_v < C$, each voice user transmits in a request slot with probability 1 and each data users transmits with probability $p_d = (C - N_v)/N_d$. Therefore,

$$p_s(N_v, N_d) = \sum_{i=0}^{N_d} p_{i+N_v} P(i \text{ data users transmit in this frame})$$

$$= \left(1 - \frac{1}{M}\right)^{N_v - 1} \left(1 - \frac{p_d}{M}\right)^{N_d}.$$

The throughput consists of successfully transmitted voice and data requests:

$$T(N_v, N_d) = \sum_{i=0}^{N_d} T_{i+N_v} P(i \text{ data users transmit in this frame})$$

$$= N_v \left(1 - \frac{1}{M}\right)^{N_v - 1} \left(1 - \frac{p_d}{M}\right)^{N_d}$$

$$+ (C - N_v) \left(1 - \frac{1}{M}\right)^{N_v} \left(1 - \frac{p_d}{M}\right)^{N_d - 1}. \tag{5}$$

3. When $N_v = 0$, data users transmit with probability p_d, $p_d = \min(1, M/N_d)$, to maximize the throughput.

$$T(0, N_d) = N_d p_d \left(1 - \frac{p_d}{M}\right)^{N_d - 1}. \tag{6}$$

3.3 Stability Analysis

We now prove that our algorithm is stable with a fairly weak assumption on the arrival process. We consider a system with a unique stationary distribution as a stable system. We use Pake's Lemma to find a sufficient condition for the system to be stable [9].

Lemma 1 (Pake's Lemma). *Let $\{X_k, k = 0, 1, 2, \cdots\}$ be an irreducible, aperiodic homogeneous Markov chain with state space $\{0, 1, 2, \cdots\}$. The following two conditions are sufficient for the Markov chain to be ergodic.*

a) $|E(X_{k+1} - X_k | X_k = i)| < \infty, \ \forall \ i,$

b) $\limsup_{i \to \infty} E(X_{k+1} - X_k | X_k = i) < 0.$

Note that an irreducible, aperiodic, ergodic Markov chain has a unique stationary distribution.

Let A_k be the total number of users that arrive in the kth frame. Suppose that $\{A_k, k = 0, 1, 2, \cdots\}$ are random variables with mean value λ. Let X_k be the number of users (voice users and data users) at the beginning of the kth frame, then $X_k = N_v + N_d$. Let $B(X_k)$ be the number of users whose requests are successfully transmitted in the kth frame. We now prove that $\{X_k, k = 0, 1, 2, \cdots\}$ is ergodic using Pake's lemma. We have

$$X_{k+1} = X_k + A_k - B(X_k).$$

So, for any i,

$$|E(X_{k+1} - X_k | X_k = i)| = |E(A_k - B(X_k) | X_k = i)|$$
$$= |E(A) - E[B(i)]| \leq |E(A)| + |E[B(i)]| \leq \lambda + M < \infty.$$

Hence, condition (a) of Pake's lemma is satisfied.

To satisfy condition (b) of Pake's lemma, we require that

$$\limsup_{i \to \infty} E(X_{k+1} - X_k | X_k = i)$$
$$= \limsup_{i \to \infty} E(A_k - B(X_k) | X_k = i)$$
$$= \limsup_{i \to \infty} (\lambda - E[B(i)]) < 0.$$

So

$$\lambda \leq \liminf_{i \to \infty} E[B(i)] \tag{7}$$

is a sufficient condition for the system to be stable.

In our QoS algorithm, when there are N_v voice users and N_d data users, the total number of users is $i = N_v + N_d$. Then, there exists an L such that for all $i \geq L$, we have [6]:

$$T(N_v, N_d) \geq C \left(1 - \frac{C}{iM}\right)^{i-1}.$$

Hence,

$$B(i) \geq C \left(1 - \frac{C}{iM}\right)^{i-1}.$$

Then

$$\liminf_{i \to \infty} E[B(i)] \geq \liminf_{i \to \infty} C \left(1 - \frac{C}{iM}\right)^{i-1} = Ce^{-\frac{C}{M}}. \tag{8}$$

Hence, from (8), $\lambda \leq Ce^{-C/M}$ is the sufficient condition for the system to be stable under the QoS requirement $\bar{P}_s \geq A_0$, where λ is the arrival rate. Note that in the special case $C = M$; i.e., the system is designed to achieve the maximum achievable throughput, (8) becomes:

$$\liminf_{i \to \infty} E[B(i)] = \liminf_{i \to \infty} M \left(1 - \frac{1}{i} \right)^{i-1} = Me^{-1} \tag{9}$$

The sufficient stable condition is $\lambda < Me^{-1}$, which is exactly the stable condition for slotted ALOHA. Furthermore, there is no bistable point in the system because the throughput does not decrease when the number of blocked users in the system increases.

3.4 Upper Bound on Throughput

We consider the QoS requirement as $\bar{P}_s \geq A_0$. With this restriction, we derive an upper-bound on the throughput for random access algorithms satisfying the following two assumptions. First, all users transmit in request slots randomly and independently. Second, each user transmits in at most one request slot in each frame. Let Ω be the set of all such random access algorithms.

We consider the throughput under two conditions. Condition 1: there is at least one voice user in the system. Condition 2: there is no voice user in the system. First, we consider the throughput under Condition 1. Let X denote the total number of users that transmit in this frame, $1 \leq X \leq \infty$. The probability that the voice user successfully transmits its request in this frame is p.

$$p = \begin{cases} p_X & : \quad \text{if the user transmits in this frame,} \\ 0 & : \quad \text{otherwise,} \end{cases}$$

where

$$p_X := \left(1 - \frac{1}{M} \right)^{(X-1)}.$$

Note that

$$E(p_X) = E\left(\left(1 - \frac{1}{M} \right)^{(X-1)} \right) \geq E(p) = \bar{P}_s \geq A_0. \tag{10}$$

Let T_1 be the throughput given that there is at least one voice user in the system. Then,

$$T_1 = E\left(X \left(1 - \frac{1}{M} \right)^{(X-1)} \right). \tag{11}$$

We want to maximize (11) with the constraint (10). Let $Y = (1 - 1/M)^{(X-1)}$. So

$$E(Y) \geq \bar{P}_s = A_0 = \left(1 - \frac{1}{M} \right)^{K_0 - 1}.$$

Let $f(y) = -y \left(\frac{\ln y}{\ln \left(1 - \frac{1}{M}\right)} + 1 \right)$, which is a strictly convex function. By Jensen's inequality [1],

$$T_1 = E\left(-f(Y)\right) \leq -f(E(Y)) = K_0 \left(1 - \frac{1}{M}\right)^{K_0 - 1} =: T_C. \tag{12}$$

Next, we consider the condition 2; i.e., no voice user is in the system. Let T_0^α be the throughput of a random access algorithm α when there is no voice user in the system. Let $T_0^m = \max\{T_0^\alpha, \ \alpha \in \Omega\}$. Let q_α denote the probability that no voice user is in the system of a random access algorithm α. Let $P_0 = \max\{q_\alpha, \ \alpha \in \Omega\}$. For algorithm α, let P_1^α be the probability that there is at least one voice user in the system. Hence, $1 - P_1^\alpha \leq P_0$. The throughput T of algorithm α is given by:

$$T = T_1 P_1^\alpha + T_0^\alpha (1 - P_1^\alpha) \leq T_C P_1^\alpha + T_0^m (1 - P_1^\alpha)$$
$$= T_C + (1 - P_1^\alpha)(T_0^m - T_C) \leq T_C + P_0 (T_0^m - T_C) =: T_{max}. \tag{13}$$

Therefore, T_{max} is the upper-bound on the throughput of random access algorithms in Ω (algorithms such that all users transmit for request slots randomly and independently, and each user transmits for at most one time slot in a frame). This upper-bound is not restricted to the (p_v, p_d) strategy used in this paper.

The above upper-bound, T_{max}, may not be tight. We compare T_1 with T_C. Since f is a strictly convex function, (12) achieves equality when $Y = E(Y)$ with probability 1. Hence, the upper-bound T_C is only achievable if $X = C$ with probability 1; i.e., there are always exactly C users transmitting in each frame. However, C may not be an integer and it may not be possible to let exactly C users transmit in random access algorithms. So T_{max} may not a tight upper-bound. We try to approach the upper-bound by assigning p_v and p_d such that $E(N_v p_v + N_d p_d) = C$ in our scheme.

Next, we show that

$$\lim_{M \to \infty} \frac{T(N_v, N_d)}{T_{max}} = 1,$$

when there are enough users in the system.

With some tedious algebra [6], we can show that

$$T(N_v, N_d) \geq T_C \left(1 - \frac{1}{M}\right),$$

when

$$N_v + N_d \geq C.$$

Recall that P_0 is the maximum probability that there is no data user in the system. Let p_0 be the probability that there is no new voice user with a request in a frame. Then, $P_0 = p_0 P(\text{all voice users with requests transmit successfully by the end of a frame in steady state})$. In practice, P_0 is small when M is large; i.e., in a large frame, it is unlikely there is no voice user in the frame. For example, if the arrival process of voice users is a Poisson process with mean vM, then $P_0 \leq p_0 = e^{-vM}$. Suppose $P_0 \rightarrow 0$ as $M \rightarrow \infty$. We have

$$\frac{T(N_v, N_d)}{T_{max}} \geq \frac{T_C \left(1 - \frac{1}{M}\right)}{T_C + P_0(T_0^m - T_C)} \rightarrow 1.$$

Hence, the throughput of the presented QoS algorithm asymptotically approaches the upper-bound. In other words, when there is at least one voice user in the system, the throughput of our QoS algorithm approaches T_C. Furthermore, as M goes large, the probability that there is no voice user in the system goes to zero. So the throughput of our QoS algorithm asymptotically approaches the upper-bound.

4 Simulation Results

In this section, we provide simulation results of the proposed scheme. For all simulations in this section, we set $M = 20$; i.e., there are 20 request slots in each frame. For each figure, we run simulation for $100,000$ frames in a single-cell. We assume the arrival processes of voice users and data users are independent Poisson processes with the same average rate.

At the beginning of each frame, the base station announces N_v and N_d, the numbers of voice users and data users in the system. (The announced numbers are estimated by the base station in the practical approach.) Knowing N_v and N_d, each user decides its transmission probability according to (2). With this probability, the user selects and transmits in a request slot in the frame. If the user is the only one transmitting in its request slot, its transmission is successful. Otherwise, the user has to wait for the next frame to retransmit and its delay is increased by one. The unit of delay is frame.

Figure 1 indicates the delay distribution of a voice user when $A_0 = 0.6$, where A_0 is the required success probability. We can see that the delay distribution of a voice user is well approximated by a geometric distribution when the numbers of new arrived users at different frames are independent and $M = 20$. Hence, in other figures, we use the success probability as the delay performance measure.

Figures 2 and 3 illustrate the performance of the proposed QoS algorithm. Figure 2 indicates the delay performance of voice users. The delay performance is shown by the average probability of success. Simulations are run under both the ideal condition and the practical condition. By the ideal condition, we mean that the base station knows the exact numbers of voice and data users in the system. In practice, a Kalman filter is used to estimate the numbers of users. The Kalman filter approach is implemented with two threshold values. We use (3) to approximate C. In the ideal condition, (3) offers a pretty good approximation. With $C = K_0$ in the Kalman filter approach, $\bar{P}_s$ is less than the QoS requirement due to estimation errors. Thus, in practice, we should use a smaller threshold value than the one calculated under the ideal condition, which is represented by the curve with $C = 0.9K_0$. Figure 3 shows the throughput performance. It is

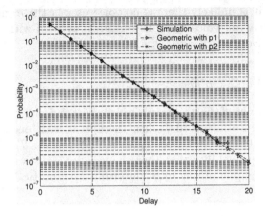

Fig. 1. Delay distribution of a voice user when $M = 20$, $p1$ is the reciprocal of the average delay of voice users, and $p2$ is the average probability of success of voice users.

obvious that the throughput decreases with the increase of A_0. We compare the throughput in the ideal condition with the practical approaches. As expected, the Kalman filter approach with the smaller C has less throughput, illustrating the tradeoff between the throughput and QoS. We use the probability of no new voice user in a frame, $p_0 = e^{-r\lambda}$, as the upper-bound of P_0, where P_0 is the probability of no voice user in a frame. Hence, p_0 is used to calculate the upper-bound of throughput shown in Figure 3, which results a looser upper-bound than that in (13). However, we still note that in most cases, the throughput in the ideal condition is quiet close to the upper-bound in the figure.

5   Conclusions

We present a random access scheme that provides certain QoS guarantees during the contention phase of communication. Permission probabilities are used to provide QoS for two traffic classes, voice users and data users. The same idea can be extended to multi-class users. The QoS requirement of voice users is defined as $\bar{P}_s$, the average success probability of voice users. For a predetermined QoS measure $\bar{P}_s$, a threshold C is calculated such that a voice users has an average success probability larger or equal to $\bar{P}_s$. We prove that the algorithm is stable with a weak assumption. We derive the upper-bound of a general class of random access algorithms under the QoS requirement in term of $\bar{P}_s$ and show that the studied algorithm asymptotically approaches the upper-bound. The analysis is based on the QoS algorithm without capture. Note that the QoS algorithms with and without capture are the same in essence except that the success probability is higher when capture is considered [6].

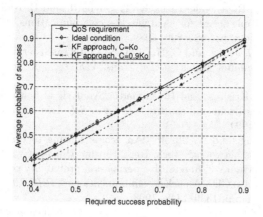

Fig. 2. Delay performance without capture for $M = 20$ with 50% voice users. In the legend, KF denotes Kalman filter.

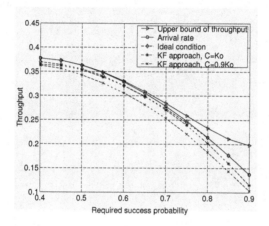

Fig. 3. Throughput without capture for $M = 20$ with 50% voice users. In the legend, KF denotes Kalman filter.

In wireless networks, providing QoS during contention phase is important to support bursty traffic. It is quite different from the wire-line scenario. So existing methods such as using in ATM do not apply directly. There would be large research space for this topic.

References

[1] P. Billingsley, *Probability and Measure*. Wiley, 1985.
[2] C. Bisdikian, "A review of random access algorithms," *IBM Res. Rep.*, no. RC20348, Jan. 1996.

[3] D. J. Goodman, R. A. Valenzuela, K. T. Gayliard, and B. Ramamurthi, "Packet reservation multiple access for local wireless communications," *IEEE Trans. Commun.*, vol. 37, no. 8, pp. 885–890, 1989.

[4] W. S. Jeon, D. G. Jeong, and C.-H. Choi, "An integrated service MAC protocol for local wireless communications," *IEEE Trans. Veh. Technol.*, vol. 47, no. 1, pp. 352–363, 1998.

[5] R. LaMaire, A. Krishna, and H. Ahmadi, "Analysis of a wireless MAC protocol with client-server traffic and capture," *IEEE J. Sel. Areas Commun.*, vol. 12, no. 8, pp. 1299–1313, 1994.

[6] X. Liu, E. Chong, and N. Shroff, "An access scheme to provide qos in packet-Switched wireless networks," Tech. Rep., Purdue University, 2000.

[7] K. Mori and K. Ogura, "An adaptive permission probability control method for integrated voice/data CDMA packet communications," *IEICE Trans. Fundamentals*, vol. E81-A, no. 7, pp. 1339–1348, 1998.

[8] H. Peyravi, "Medium access control protocols performance in satellite communications," *IEEE Communications Magazine*, vol. 37, no. 3, pp. 62–71, 1999.

[9] R. Rom and M. Sidi, *Multiple access protocols : performance and analysis.* New York : Springer-Verlag, 1990.

Mobility Management and Roaming with Mobile Agents

Do Van Thanh[1], Sverre Steensen[2], and Jan A. Audestad[3]

[1] Ericsson Norway Applied Research Center, P.O box 34, N-1361 Billingstad,
Norway, Tel: + 47 66 84 12 00
etodvt@eto.ericsson.se
[2] University of Oslo - Unik, P.O box 70, N-2007 Kjeller,
Norway, Tel: + 47 63 81 45 70
sverrest@ifi.uio.no
[3] Telenor AS, P.O box 6701, N-0130 Oslo, Norway, Tel: + 47 22 77 99 52
Jan.Audestad@telenor.no

Abstract. In this paper we propose that the mobile agent concept can be used in the design and implementation of ideal user mobility. The paper starts with a discussion of the traditional mobility concepts; terminal and personal mobility. The concept of ideal user mobility is introduced which can be seen as an extension of these two concepts better suited for the needs of the mobile user. Then, the fundamental mechanisms for mobility support are described and how they can be implemented by mobile agents. Strategically mobile agents which can combine all these mechanisms in a flexible way is then discussed. The paper concludes with a summary of why mobile agents are well suited to provide mobility support and some suggestions for future studies.

Background:

The work described in this paper is a result from the mobility part of the TELEcom REsearch Program TELEREP, carried out at the Ericsson Norway Applied Research Center (NorARC) in collaboration with the University of Oslo, Center of Technology at Kjeller Unik [1]. The program has been established to obtain practical experience with Distributed Object Computing methods such as ODP/TINA [2], [3], [4], CORBA[5], mobile code in the implementation of TMN and IN functionality, and to study how mobility and security can be provided in the DPE environment.

1 Introduction

With the growth of the computer industry computers are showing up everywhere. Today most people interact with a number of different computers daily, at their job, at home or when travelling. All these computer users know that taking the mobile computer on the road or accessing a remote computer is something completely different than doing work from their normal environment. In other

C.G. Omidyar (Ed.): MWCN 2000, LNCS 1818, pp. 123–137, 2000.

words, the situation for the mobile users are far from being ideal. Problems like missing applications, unfamiliar user interfaces and unavailable services are normal. A user changing terminals and network access points is most often facing a frustrating and time consuming task to access services that are normally used.

There has been much optimism surrounding the concept of mobile agents. Mobile agents like the users move around in heterogeneous networks. With a *mobile agent* we mean a computer program that acts autonomously on behalf of a person or organization, and is not bound to the system where it began execution, as defined by the MASIF standard [6].

It is the intention of this paper to combine the user's need for mobility with mobile agent technology witch has great promise and built in mobility concepts that has not yet been realized in large commercial applications. In order to do so we will first examine general mobility concepts, before the fundamental mobility mechanisms used to implement these mechanisms are presented. Then the concept of strategically mobile agents, and how each of the mobility mechanisms can be supported by agents, are discussed.

2 Mobility Concepts

This section will present the concept of terminal and personal mobility, which can be achieved independently of the technology used for the implementation. However, implementing these mobility concepts are not enough to provide user friendly mobility services. We therefore propose the concept of ideal user mobility to extend terminal and personal mobility.

Terminal Mobility
Terminal mobility is the ability to provide service to a terminal that moves around, or is connected to a variety of network access points (NAPs). To be able to access the network the terminal has to be connected to a NAP. To provide network access there must exist an association between the terminal and the NAP. This association has to be dynamic to allow for terminal mobility. The NAP has physical characteristics that restrict the kind of terminals that can use the NAP and the type of services that can be provided. In telecommunication, terminal mobility has been supported by keeping the user's current address continuously updated at a database located at the user's home domain, as is the case for Mobile IP and GSM. This allows for terminal mobility, but the services that the users can access have been very limited. The computing industry has had more trouble supporting terminal mobility. On a large part the management of addresses has been left to the users themselves, which limits the usefulness of terminal mobility.

Personal Mobility
TINA-C defines personal mobility as the ability to let the user access services from any terminal at any location [7]. ETSI takes this definition a little further

and states that access to a terminal is based on a personal telecommunication identifier, and the network must provide services according to the user's service profile [8]. We will use the ETSI definition in this paper. To be able to provide services to a user connected to a terminal on a network there must exist an association between the user and the terminal. This association must be dynamic in order to ensure personal mobility. A user might have a profile. This profile records the user's preferences and there must exist some compatibility between this profile and the capabilities of the terminal. For instance if the user requires a video session, but the terminal can only offer voice capabilities or services, some compromise must be made, or the terminal must reject the request for service. What personal mobility really ensures is that the user can access a set of basic services across the network from a number of different terminals connected to a variety of NAPs. However, these basic services are not enough for most users. They also need to have access to their customized applications, data and profile. The term personal mobility does not ensure this and it is up to the users themselves to expand the services offered.

Ideal User Mobility
Even when terminal and personal mobility are combined the mobile user might still not be satisfied. We propose the concept of ideal user mobility which incorporates the concepts of terminal and personal mobility while adding transparency on top of these concepts and extending the number of applications, data and profile that can be accessed. Ideally the different computing environments should adapt to the users instead of the users having to adapt to them. *Ideal user mobility* can then be seen as the ability to provide the same applications, data and profile transparently to the user anywhere, anytime and with the same look and feel. The concept of ideal user mobility used in this paper has some analogies to the UMTS's Virtual Home Environment (VHE) concept. It is defined as a *system concept for personalized service portability across network boundaries and between terminals* [9], [10]. However, the VHE concept focuses only on service while our concept of ideal user mobility comprises of services/applications, data and profile.

Some alterations to ideal user mobility can be accepted if the terminal used to access the network clearly does not have the capabilities needed to provide the requested services. By capabilities we mean limitations in memory, processing power, display capabilities etc. If this is the case a negotiation between the user and the terminal should decide what services can be offered. For instance if a user accesses the computer network through a cellular telephone, the obvious limitations in display and computing capabilities restrict the available applications that can be accessed, and the degree of ideal user mobility that can be achieved. Such an alteration will be more acceptable to the user because of the obvious limitations of the terminal.

3 Fundamental Mobility Mechanisms

When the user is away from the home domain at a visiting domain there are several alternatives to enable access to the applications, data and profile. After studying systems offering mobility like GSM, UPT, Mobile IP and Telnet, the fundamental mechanisms used to provide mobility are classified into two categories:

Establishment of a Channel from the Visiting to the Home Domain

This is a fairly simple approach which resembles how traditional telephony works. A channel is established between the visiting and home domain and data is transported over the channel. This can possibly give the user access to synchronously and asynchronously communication applications, computational applications, user data, and the profile stored at the home domain.

1. *Continuously open channel:* A continuously open channel is a channel that after it has been established, remains open whether it is used or not. The user is then guaranteed to have a certain amount of bandwidth available at all times, whether the channel is used or not. This approach is adopted by for instance Telnet, where a channel is kept open after the two Network Virtual Terminals have negotiated the available services. The channel is not closed until the user requests it, or some network failure occurs. This approach is well suited if the user needs quick access to his home environment, and a fairly stable connection can be established between the two domains. This mechanism is best suited for synchronous communication applications. The disadvantages of this approach are that it can be very cost inefficient to keep an open channel between the home and visiting domain. In addition security restrictions on both sides can restrict the environment that can be accessed through the open channel. If the channel has low capacity for data transmission (which is often the case), the open channel approach can be too slow, especially if the user requires fast interaction with applications on his home domain.

2. *Discontinuously open channel:* Instead of keeping a channel continuously open, a channel can be established only when the user is about to send or receive data. After the final bit of the transmission has been sent the channel closes. Actually such a discontinuously open channel is only a virtual channel, and transmission is usually based on "best effort" delivery of data packages. Mobile IP employ this approach. The use of a discontinuously open channel is well suited if the user only needs access to his home domain for limited periods, or if he wants to be notified and receive incoming data when it is sent to his home domain. This approach is best suited to support asynchronously communication applications. This approach faces problems from low quality network connection, and possibly slow interaction with home applications. By its nature the discontinuously open channel approach is not well suited to support synchronous communication applications since transmission is normally based on best effort.

Duplication of Applications, Data, and Profile

This approach has been extensively used by both the telecommunication and computing industry. Traditionally applications, data files and profile have been moved between computers on storage mediums before they were installed on a new computer. This was a tedious task requiring extensive knowledge from the users. The duplication mechanism can be split into two:

1. *Pre-duplication:* With pre-duplication we mean that the applications, user data and profile are copied and installed at the visiting domain before the user arrives. Pre-duplication can again be split into two:

   (a) *Static pre-duplication:* When applications, data and profile are copied to the visiting domain before the user arrives and remains after the user has logged off, we have static pre-duplication. In GSM the Basic service and some of the supplementary services are statically pre-duplicated. These services are standardized and are identical on all hosts. Advantages of this approach are that the applications that are duplicated this way, will be quickly available to the users when they register at the visiting domain. The users can also rely on these particular services being available almost everywhere. This approach is best fitted for supporting applications that are used by a broad range of users, and are large and complicated. Disadvantages of this approach are that they are not very well suited for giving access to user data or profile. Nobody wants to have their files spread all over the network, and inconsistency between all these files could become a large problem. Standardization is often required to achieve effective static pre-duplication, but this process is normally time consuming and will increase a new service's time to market. Changing existing services is difficult since applications are distributed on many different hosts.

   (b) *Dynamic pre-duplication:* If the user specifies where he is going and when, his applications, user data and profile can be transported to that domain ahead of him. When the user leaves the visiting domain the applications, data and profile are destroyed or transported back to his home domain. We do not know of any systems currently supporting this form of duplication. Advantages of this approach are that the applications, data and profile are transported ahead of the user. This ensures a quick initialization process when the user logs in. When logged in the user can interact fast and locally with his application. This approach is best suited for potentially large applications that the user needs access to but will not wait for while it downloads. A disadvantage of this approach is that it requires security mechanisms to be in place so that the user's data and profile will not be spread around in the network.

2. *Dynamic duplication:* Dynamic duplication takes place after the user has registered at the remote host. It is a process where applications, user data and profile are copied from the user's home domain to the visiting domain. This form of duplication requires that there is some sort of platform interoperability between the two domains, and that there can be allocated enough

resources at the visiting domain. GSM uses this approach to move the user profile from the home to the visiting domain. Advantages of this approach are that it might be an excellent alternative if the user is going to access the network over a long period. It is then very cost efficient, while providing fast interaction with applications and data. Dynamic duplication is well suited for computational applications. Disadvantages of this approach are that it might be impractical to use if the amount of data to be transfered is very large. It would then require extensive transmission which can be a time consuming process. It does not support communication applications well.

It is not possible to conclude that any one of these mechanisms are superior to the others. They all have their advantages and disadvantages and are well suited for some situations and applications but not for others. Obviously a combination of these mechanisms will offer the best solution. This combination should be flexible adapting to the user's needs and situation, in order to provide ideal user mobility. Any concept that permits a dynamic and flexible combination of these mechanisms will be suitable for the realization of ideal user mobility.

4 Mobile Agents Supporting Mobility

So far very little research has been done on how mobile agents can support mobile users. Most mobile agent applications suggested have been based on fixed users sending out mobile agents to accomplish tasks like information searching and electronic shopping. Gray et al. have suggested that mobile agents can offer advantages to the mobile user [11]. They focus on mobile users that are partially connected through often unreliable network connections with laptops or personal digital assistants (PDAs). In our terminology they are investigating ways to support terminal mobility and their focus has been on how to make efficient use of network resources. Through experiments they demonstrated that mobile agents can leave laptops and return despite disconnection and reconnection at different IP-addresses. The purpose of this paper is to see how mobile agents can implement the mobility mechanisms discussed in the previous section. Before these implementations are presented the most central agents are presented briefly:

Domain Agent(DA)
At every domain supporting our implementation there should exist one DA. It is created before any other agents can populate the domain, and is a stationary agent remaining at the same host for its lifetime. Its main responsibility is to manage the domain's resources and enforce security restrictions.

User Agent (UA)
The UA is the user's main representative in the system. It could be created when the user subscribes to the system, or by the user through an agent generator program. The UA is responsible for providing services to the user based on the user's specifications. It needs to communicate with the user, the DA and

the other agents and should therefore have advanced communication abilities. When it migrates it clones itself so a copy is always left on the home domain. This ensures that the UA is not lost or destroyed completely if things go wrong during migration.

User Data Agent (UDA)
The UDA is responsible for collecting the user's data files according to the UA's specifications. These files can be spread at the user's home domain and are copied and collected when the UDA is sent for. The reason for only dispatching copies of the files instead of the original data files are two fold: First, if something goes wrong during transfer the data can be permanently lost. Secondly if only a copy is sent, it would be unnecessary to return data files that are unchanged. A better solution is just to destroy those files remotely. The UA therefore only contains the logic necessary to collect those files, not the files themselves.

User Application Agent (UAA)
The UAA is like the UDA created by the user or UA, and remains de-activated until the UA retracts it. The UAA contains logic or program code, making it able to collect the necessary files to provide the applications specified by the UA.

User Profile Agent (UPA)
The UPA is created by the user or UA, and as the other task agents remains de-activated until it is retracted by the UA. The UPA is responsible for collecting the customizations that the user has done to the applications and the subscribed services the UA has asked for. There would be no point in bringing the user's entire profile over if only some services and applications are needed. The UPA should therefore collect that part of the profile that is best adapted to that particular domain.

Mobile Agents Supporting Dynamic Duplication
We will now present how mobile agents can be used to implement dynamic duplication. Our basic model can of course be extended to offer more sophisticated services but we believe that this model presents the core concepts. We have presented this model in figure 1. To get a better understanding on how dynamic duplication can be implemented we will go through the model based on the numbers in the figure.

1. When a user logs in at a terminal connected to the network the DA presents him with a graphical user interface (GUI). At the GUI the user should provide his user identifier and password. In order to support a transparent login scheme the user ID needs to be valid at all hosts supporting the agent system. The identifier can be a combination of the user's domain address and a user identifier, or an independent identifier not based on the user's home domain's address. When the DA knows the user's home domain it retracts the user's UA.

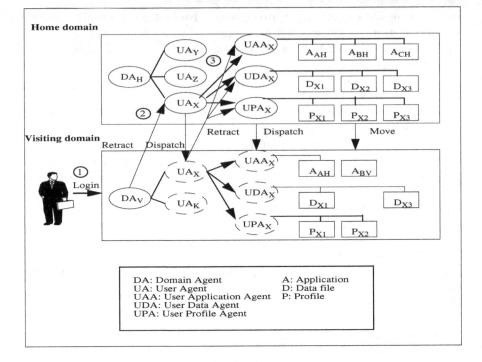

Fig. 1. Mobile agents supporting dynamic duplication

2. When the UA receives the retract call it should migrate to where the migration call originated. After arriving at the remote host it registers with the DA. At this point the quality of the user's session should be negotiated including access rights to resources, allocated memory, CPU cycles and physical memory. The UA then presents these restrictions to the user along with a GUI. From the GUI the user decides what applications, data and profile he needs. The user can decide directly by naming what to bring explicitly, or more indirectly based on criteria such as: The most recently used files, files on certain formats, files belonging to certain projects etc. As we can see from figure 1, application B is already present at the visiting domain so there is no need to bring over that application. The user does not need application C, so it is not brought over. When the user has finished specifying what is needed, the UA is responsible for retracting the specified agents.

3. Agents receiving the retract call collect the specified files which might possibly be spread around the user's home domain. The reason for not collecting all the files, when the agents are initialized, is that the file's content might change in the meantime. These files are loaded into the agent's memory before the agents dispatch to the visiting host. When arriving at the visiting host, the agents write their data to files and distribute them in a way so that the user can have easy access to them. The applications, data and profile are now available to the user.

When the user decides to log off he does so by clicking on the log off button on the GUI presented by the DA. The DA then sends a dispose call to the UA, which sends dispose messages to the other agents. When receiving the dispose call the agents examine the files they brought over to see if they have changed. If they have, they are transported back to the user's home domain by the appropriate agent. If no changes have occurred the agents are garbage collected at the visiting domain. After the application, data and profile agents have been disposed the UA is disposed and garbage collected.

The dynamic duplication mechanism gives the user access to his computing applications, data and profile. The user can get the communication applications duplicated as well, but it would only be useful for outgoing communication. Incoming communication would still be routed to the user's home domain and duplicating the applications will not be sufficient to receive communication. However, it takes time to dynamically duplicate which may be frustrating to the user.

Mobile Agents Supporting Dynamic Pre-duplication

In the dynamic duplication process the user has to wait while the environment is being initialized. This process could possibly take a long time, depending on the quality of the communication channel and the size of the data being transported. If the user knows where he is going and at what time he will be arriving at a certain domain, the pre-duplication mechanism can be employed. His agents can be sent ahead of him initializing his environment and preparing the domain for his arrival. When the user then logs in, his environment will be initialized quickly. We have not illustrated the design of this scenario since it is based on the dynamic duplication model.

When the user is at his home domain knowing that he will need to access a visiting domain later, he can have his agents migrate before him. In order to achieve this he needs to specify the visiting domain's address and when he will be there. This could be done each time the user travels, or the UA could have access to the user's time manager and follow that timetable. When the user is deciding on what applications, data and profile to bring, there should at least exist two options:

1. The user can specify exactly what applications, data and profile he wants access to. The UA then migrates to the host and negotiates with the DA to achieve the best possible environment for the user. After the negotiation has taken place the UA retracts the other agents. The rest of the process is similar to dynamic duplication.
2. Instead of the user having to specify exactly what to bring over, the UA can be responsible for negotiating the best possible environment based on a pre-defined strategy. The agents are then retracted in the same way as in dynamic duplication.

When the user arrives and logs in, his environment is presented. If the user somehow does not show up, the UA should have a timer that times out resulting in garbage collecting of all files brought over. This would prevent against the user's files being spread around the network. As with dynamic duplication changes to the files are transported back to the user's home domain by the appropriate agents.

As we have seen pre-duplication relieves the user of the time consuming initialization process. On the other hand it also removes some of the flexibility and adaptiveness of the dynamic duplication process, since the user must specify where and when he will be visiting the remote domain. This approach could be extended further if the user has certain patterns of movements. As with dynamic duplication it does not support access to communication applications satisfactorily.

Mobile Agents Supporting Static Pre-duplication
Static pre-duplication is the process of copying and installing applications, data and profile to a new domain and let them remain there after the user has logged off. As discussed under network management this is an often mentioned application area for mobile agents. The mobile agents could be used to install for instance certain applications at nodes in a network or between different domains. However, static pre-duplication can for all practical purposes only be used to pre-duplicate applications. Having user data and profile statically pre-duplicated will limit the user's privacy, and it will be very difficult to keep the data and profile consistent between nodes. Static pre-duplication can be based on the dynamic pre-duplication mechanism with the only difference that what the agents brought over remains at the domain after the user has logged off. After the agent has installed the applications at the domain it could return to its home domain or travel to other hosts, with the same applications.

Mobile Agents Supporting a Continuously Open Channel
A system based on mobile agents can also be used in establishing a continuously open channel between the home domain and the visiting domain. We have presented an overall model for this design in figure 2. We will now go through the numbered steps:

1. The initialization phase is identical to that of dynamic duplication. The user logs in at a remote host and with his normal user identifier and password.
2. The DA at the visiting domain then retracts the UA from the visiting domain. When the UA arrives it presents a GUI to the user with the option *Establish a continuously open channel*. The user then selects this option.
3. The UA then negotiates with the DA or possibly a Network Agent (NA). The NA is responsible for ensuring a certain quality of service (QoS) on the network connection. This can be defined through parameters such as throughput, transit delays, security of communication, priority etc. After the negotiation the UA is allowed by the DA to establish a connection to the user's home domain with the negotiated QoS. The UA located at the user's home domain is notified about the resources available at the visiting

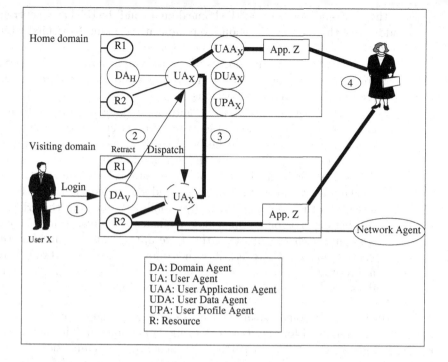

Fig. 2. Mobile agents supporting a continuously open channel

domain. This ensures that if someone wants to communicate with the user by for instance video conferencing, and those capabilities do not exist at the visiting domain, the request for communication could be denied or some compromise should be made. The channel is then established with the original UA left at the user's home domain and the migrated UA functioning as endpoints. The UA then, rather than the user itself, is responsible for maintaining the connection. If for instance the connection goes down, the UA should re-establish the connection transparent to the user. Also if the connection is characterized by bursty network traffic, the UA should be able to dynamically reserve more network resources to accommodate that traffic by negotiation with the NA.

4. If someone wants to communicate with the user the user's normal home address will be used. When the request reaches the user's home domain it is received by the appropriate application which forwards the communication to the UAA. It again contacts the UA. When receiving the request for communication the UA should consider the following:

   (a) If the user is logged in at his home location and the resources needed are available, communication is routed to the appropriate resource and communication can be engaged.

(b) If the user is not logged in at the home domain and the UA has no current address of the user, synchronous communication should be denied while asynchronous communication can be accepted.

(c) If the UA has the current address of the user and there is established a continuously open channel between the home and visiting domain, the UA should at least have two options:

    i. It could return the user's current address. In doing so a direct channel can be established between the two parties, circumventing the user's home domain. For this to be possible there has to exist necessary applications and resources at the visiting domain.

    ii. The UA at the home domain could also establish a channel on behalf of the user between the requester and the home domain. This is similar to the approach taken by Mobile IP, but here the triangle routing depends on an open channel instead of a discontinuous one. This approach requires that there are appropriate resources at the visiting domain. However, since the communication goes through the appropriate application at the home domain, similar applications does not have to exist at the visiting domain.

The UA on the home domain can then be seen as responsible for forwarding communication intended for the user sent to his home domain. The UA at the visiting domain is responsible for sorting the incoming communication according to what application type it belongs to and the resources it needs (R). For instance if the user is talking to someone on an IP telephone and he wants to establish a text based *talk* session with someone else, the UA will be responsible for splitting incoming data to its appropriate resource. In figure 2 we see that the User Agent routes the incoming data to resource *R2*.

This approach gives the user access to his computational and communication applications but in a much slower fashion than if it was accessed locally. Continuous interaction can then become a tedious and frustrating process. This approach can give the user access to his applications, data and profile but at the cost of keeping an open channel and slow interaction with applications at the user's home domain.

Mobile Agents Supporting a Discontinuously Open Channel
A discontinuously open channel can also be supported by mobile agents. This approach is based on the way mobile IP works, but while mobile IP only provides the user with incoming data, this will in addition provide the user with access to his home environment. We have not illustrated this mechanism because it closely resembles the continuously open channel approach.

When the user arrives at the remote host he retracts his user agent as usual. When the UA arrives it presents a GUI to the user and a discontinuously open channel is decided on. The UA then sends a message back to the UA at the home domain notifying it of its choice. The user can then start working at the host. When computational applications, data or profile are needed he indicates it to the UA who opens a channel to the home domain. This channel is shut down when the user decides to do so, or if the connection has been inactive over a longer period. If someone wants to communicate with the user, either synchronously or asynchronously, a channel is automatically established so the communication can proceed. When the user decides to log off, the UA is garbage collected.

As for the open channel approach, the user has access to computational and communication applications, data and profile. But the same problems of delays can be experienced due to remote access. However, the cost of keeping an open channel is reduced.

Combining All the Fundamental Mobility Mechanisms

We have seen that the mobile agent concept is capable of supporting the five fundamental mobility mechanisms. However, the strongest feature of mobile agents is the ability to provide a combination of all five mechanisms. A user might need:

- One or more synchronous communication applications.
- One or more asynchronous communication applications.
- One or more computational application.
- Data and profile.

In such a case all the fundamental mobility mechanisms should be employed to provide the best possible solution for the user. To illustrate how these mechanisms can be combined we consider a user moving to a visiting domain:

- Standard computational applications may have been statically pre-duplicated. The UA discovers this when arriving at the visiting host and there is no need to dynamically duplicate those applications.
- Synchronous communication applications can be moved or left at the user's home domain depending on resources present at the visiting domain and the quality of a possible continuously open channel.
- Asynchronously communication applications can also be left at the home domain and can be accessed through a discontinuously open channel.

5 Strategically Mobile Agents

As we have illustrated in the previous sections the different mobility mechanisms have their strengths and weaknesses and combining them would obviously provide best mobility support for the users. However, what constitutes such a combination for providing *ideal user mobility* is very dependent on the situation and the user's preferences. It is therefore essential to device a strategy that combines these mechanisms, in order for the agents to provide ideal user mobility. We define strategically mobile agents as agents optimizing their performance on tasks according to some strategy. This strategy can consists of:

Decision heuristics: Decision heuristics are the decision rules the agent constitutes when deciding on what mechanisms to employ. These rules are specified by the agent's owner in combination with general knowledge about distributed computing and must be decided on before the agent migrates.

Task knowledge: It is not enough to only have static decision rules on how to mix the mobility mechanisms. This would result in very little flexibility for the users. Information about the specific tasks that the user wishes to accomplish is needed to provide a flexible strategy.

Environment knowledge: The last piece of information needed to fully exploit strategic mobility is to have knowledge about the surrounding environment. No matter how good a strategy is, it will not make sense if it is not adopted to the environment. The environment poses certain restrictions on the execution of mobile agents. These restrictions can be split into *static* and *dynamic* restrictions. The *static restrictions* are initial restrictions put on mobile agents before they are allowed to execute. These restrictions are valid as long as the agent is executing at the host. *Dynamic restrictions* on the other hand change over time, according to what happens in the surrounding environment.

6 Conclusion

We believe that mobile agents, in addition to supporting terminal and personal mobility, can provide ideal user mobility. Although we have not found any research on this topic we believe that agent technology provides a useful framework that can be applied in this respect. The strongest arguments for employing agent technology to support ideal user mobility are:

- **Built in support for mobility:** Mobile agents are by their very nature mobile and all agent platforms have built in support for security although limited, and easy to use primitives for migration. Like other distributed technologies mobile agent technology hides the specific details of the underlying network. In contrast to other technologies that also hide the location of the objects, mobile agents explicitly use the location awareness to offer better services.
- **Encapsulation of code and data:** A requirement for ideal user mobility is that the user's applications (code), data (data), profile (code and data) are provided transparent to the user independent of the user's location. Mobile agents can support this requirement since they encapsulate both code and data. This is a core characteristic of mobile agent technology and is a strong argument for using it.
- **Powerful abstraction:** The agent abstraction is powerful to use for the designers and programmers, but also for the end users. Different tasks can be delegated to appropriate agents that are responsible for carrying them out. By explicitly exploiting location awareness which is a criterion for strate-

gic mobility, it can represent a better abstraction than traditional object-oriented methodology.

- **Defined models:** When mobile agent technology matures it is likely that the fundamental models we have discussed in previous chapters will be better defined and offer an established framework to the developers. Functionality like security mechanisms, naming schemes for resources, host and agents addressing, will make it a well suited technology for a heterogeneous distributed environment. When these models are in place it will ease the development of applications built on mobile agents

Mobile agents seem to be a promising concept that has the flexibility and the adaptiveness required to meet the demands from the mobile users. However, in order to be really usable, the mobile agent concept relies on the availability of technologies such as Java chips, Jini etc. which can equalize the introduced overhead. In addition, there are several issues that also need to be resolved. One crucial issue is security. It is necessary to protect the visiting site from malicious agents and reciprocally, the agent from pirate sites. The user must also be protected so that private files are not spread around the network. Last but not least, since the mobile agent concept is still new, standards are still lacking despite the effort from OMG and FIPA. This can lead to incompatibility between the different platforms and lead to slower acceptance of mobile agent technology in general.

7 References

[1] Pål Spilling. The telecom research program telerep at Unik. UNIK Aarsrapport 1996,1996.

[2] TINA-C. Overall Concepts and Principles of TINA, February 1995.

[3] ITU-TS. Basic Reference Model of Open Distributed Processing - Part 1 Overview and guide to use the Reference Model. Rec.X901(ISO/IEC 10746-1).

[4] ITU-TS. Basic Reference Model of Open Distributed Processing - Part 2 Descriptive Model. Rec.X901(ISO/IEC 10746-2)

[5] OMG. The common Object Request Broker. Architecture and specification - Rev 2.2 February 99.

[6] Dejan Milojičić and others. MASIF: The OMG Mobile Agent System Interoperability Facility. Mobility Processes, Computers and Agents

[7] TINA-C Glossary of Terms, Version 2.0, 1997

[8] ETSI, Global Multimedia Mobility: A Standarization framework for Multimedia Mobility in the Information Society, 1996

[9] 3GPP. Universal Mobile Telecommunications Systems (UMTS); Service Aspects; Virtual Home Environment (VHE) UMTS 22.70 version 3.0.0.

[10] 3GPP. Universal Mobile Telecommunications System (UMTS); Provision of Services in UMTS - The Virtual Home Environment (highlighting release 99 requirements UMTS 22.21 version 1.0.

[11] Robert S. Gray and others (1996). Mobile Agents for Mobile Computing. Technical report, Dartmouth College.

Traffic Characteristics in Adaptive Prioritized-Handoff Control Method Considering Reattempt Calls

Noriteru Shinagawa[1], Takehiko Kobayashi[1],
Keisuke Nakano[2], and Masakazu Sengoku[2]

[1] YRP Mobile Telecommunications Key Technology Research Laboratories Co., Ltd.
YRP Center, Ichibankan, 6F 3-4 Hikari-no-oka, Yokosuka, 239-0847 Japan
{shina,koba}@yrp-ktrl.co.jp
[2] Faculty of Engineering, Niigata University
2-8050, Ikarashi, Niigata, 950-2181 Japan
{nakano,sengoku}@ie.niigata-u.ac.jp

Abstract. In a cellular mobile communications system, the call in progress will be forcibly terminated, if a circuit to the destination base station cannot be secured when a handoff is attempted. Studies have therefore been performed on methods of decreasing the percentage of forcibly terminated calls by giving priority to handoff calls when the circuits are allocated. In these studies, associated traffic characteristics have been examined based on the assumption that blocked calls simply disappear. This paper proposes an adaptive handoff priority control method that varies the number of reserved circuits for handoff based on the measured of handoff blocking rate, and evaluates traffic characteristics of the proposed method and conventional one for the cases that consider and ignore reattempt calls. It was found that the proposed method could mitigate the effects of change in average speed of mobile stations and in reattempt-call parameter such as maximum retry number.

1 Introduction

When a mobile station (MS) with a call in progress moves across cell boundary in a cellular mobile communications system, the system performs circuit switching from the base station (BS) within the current cell to the BS within the destination cell to enable uninterrupted communications. This process is called "handoff." At this time, however, if a circuit to the destination BS cannot be secured, the call will be forcibly terminated. To therefore continue communicating, the connection process must be repeated, which, of course, degrades service quality. Against this background, studies have been performed on methods of decreasing forced terminations during handoff by giving handoff calls priority. This can be achieved, for example, by securing a fixed number of circuits especially for handoff, or by having the handoff process wait until a circuit to the destination BS becomes available [1-5]. These studies have been examining traffic characteristics under the condition that calls simply disappear when either blocked or forcibly terminated due to lack of available circuits to the destination BS during connection or handoff processing. It can also be considered,

C.G. Omidyar (Ed.): MWCN 2000, LNCS 1818, pp. 138-149, 2000.
© Springer-Verlag Berlin Heidelberg 2000

though, that users will often try to reconnect (reattempt call) when blocked at a new call attempt or forcibly terminated during a call.

Furthermore, when designing and operating an actual cellular system that employs a handoff priority system by securing a portion of circuits for handoff, an appropriate number of circuits for this purpose must be selected and set at each BS according to various criteria. These include the number of circuits installed at one's own BS and at adjacent BSs and the average speed of MSs that are currently connected within the cell. In addition, considering that the ratio of pedestrians and automobiles making up MSs changes over time and that speed of MS movement varies due to traffic congestion, the system must provide a flexible response to fluctuation in handoff traffic. To meet these demands, this paper proposes a handoff priority control method that adjusts the number of handoff circuits based on the measured number of failed handoffs. And we evaluate traffic characteristics of the proposed method and conventional one that reserve a fixed number of handoff circuits for both the case that assumes reattempt calls and the case that does not.

2 Proposal for an Adaptive Prioritized-Handoff Control Method

There are basically two types of methods for giving priority to handoff calls, as follows:

(1) Perform wait processing when all BS circuits are occupied at the time of new call attempt or handoff processing, and when a circuit becomes available, give priority in connection to a queued handoff call.

(2) Denoting the number of circuits at each BS as C and the number of handoff-dedicated circuits as C_h, connect a new call requesting connection only if the number of available circuits is greater than C_h. A handoff call is given priority in connection by simply connecting it if a circuit is available.

In this paper, we examine the second method that reserves circuits for handoff calls, and propose a method that controls the number of handoff circuits in accordance with the handoff traffic load so that fluctuating handoff traffic can be flexibly handled. The control sequence of the proposed method is summarized below.

For each accommodated base station, a switch incorporates the following counters: A circuit counter (C) that stores the number of circuits installed in the base station; A handoff circuit counter (C_h) that stores the number of circuits reserved for handoff circuits; A free-circuit counter (C_f) that stores the number of free circuits; A counter (C_{nh}) that counts the number of handoff control events from adjacent cells; and A handoff-block counter (C_{nb}) that counts the number of times a circuit could not be secured at the time of handoff control.

A switch will also store threshold values B_1 and B_2 (where $B_1 > B_2$) as control criteria. On receiving a connection-control request, the switch checks the counters of the base station making the request, and if $C_f > C_h$, it executes connection control and decrements the value of C_f by one, while if $C_f \leq C_h$, it blocks the new call. On the other hand, when receiving a handoff-control request, the switch increments the value of C_{nh} by one and checks the value of C_f. If $C_f > 0$ at this time, the switch executes handoff control and decrements the value of C_f by one, while if $C_f = 0$, it performs a forced termination and increments the value of C_{nb} by one. In relation to the above, an interrupt is generated at regular intervals (T) in the switch, and on detecting the interrupt, the switch calculates handoff block rate B ($B=C_{nb}/C_{nh}$) for each accommodated base station. Then, if the calculated value of B is less than predetermined threshold B_1 and if $C_h > 0$, the value of C_h is decremented by one. Conversely, if the calculated value of B is greater than predetermined threshold B_2 and if $C_h < C$, the value of C_h is incremented by one. Finally, after completing this processing for all base stations, the switch resets C_{nh} and C_{nb} of each base station to zero. In this way, the number of circuits reserved for handoff can be automatically adjusted by taking periodic measurements of the circuit block rate at the time of handoff control.

3 Simulation Model

Simulated evaluation was performed using an endless 10-by-10-cell virtual configuration, where each cell is square-shaped with side of length L and every side is connected to the adjacent cell. The number of circuits set up in each cell was uniform at C circuits, and the call arrival interval followed an exponential distribution. When the time arrives for generating a new call, a uniform random number is used to determine the coordinates, speed v, and direction θ of the MS beginning the call so that calls come to be uniformly distributed within each cell. Here, speed and direction of the MS do not change for the duration of the call. Holding time follows an exponential distribution with an average of 120 seconds.

A newly generated call will be connected to the BS if the number of available circuits in the BS is greater than C_h. Then, if a connection cannot be made, a connection request will be made again after time t. It is assumed that the MS continues to move during this time, and that it will make the next connection request in the cell that it finds itself at that point in time. Finally, if a connection cannot be made after N tries, it is assumed that the MS gives up its attempt to make the call. The time t indicating the interval from the failed connection request to the next retry follows an exponential distribution with an average of 10 seconds.

In the event that a connected call moving at speed v and in direction θ as determined at call-generation time arrives at an adjacent cell, handoff is performed if a circuit is available at the destination BS. If there is no available circuit, the call is forcibly terminated. In the case of forced termination, a connection request will be retried after t seconds in an attempt to continue the call for the portion of holding time remaining. A reconnection attempt is treated the same as a newly generated call. If a reconnection is achieved, the call continues only for the time determined by subtracting the time already used up before disconnection from the holding time set at call generation. This simulation model that takes reattempt calls into account is shown in Fig. 1 and simulation conditions are listed in Table 1.

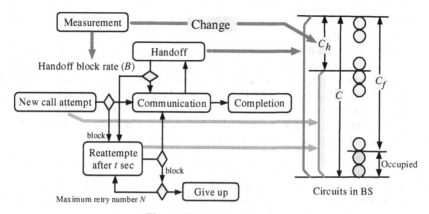

Fig. 1. Simulation model.

Table 1. Simulation conditions

Cell size	500 m
Number of circuits in cell	20
Number of reserved circuits for handoff	0, 2, 4, automatic
Speed of mobile station	0-60 km/h
Move direction of mobile station	0-2 π (uniform distribution)
Mean holding time	120 s (exponential distribution)
Mean retry interval	10 s (exponential distribution)
Maximums retry number	0, 5, 10,15, 20
Control period	600 s
B_1	0.02
B_2	0.07, 0.15

4 Evaluation Measures

When taking reattempt calls into account, calls that fully complete can be classified into the following two types: (1) calls that complete without being forcibly terminated at handoff after the call began; and (2) calls that complete after reconnecting following a forced termination at the time of handoff. In contrast, when not taking reattempt calls into account, calls that complete would fall only into the first category. Considering, therefore, that forced terminations have a significant effect on service quality, we evaluated call-completed rate R_{c1} corresponding to calls that experienced no forced termination and call-completed rate R_{c2} that includes forcibly terminated calls. These two types of call-completed rates are defined as follows.

$$R_{c1} = C_1 / C_n \tag{1}$$

$$R_{c2} = C_2 / C_n \tag{2}$$

Here, C_1 is the number of calls that completed without being forcibly terminated, C_2 is the number of calls that completed both without being forcibly terminated and after reconnecting following a forced termination, and C_n is the total number of newly generated calls.

We evaluated the "average number of connection successes" to reflect how many times a successful connection is achieved per newly generated call. A higher value for the average number of connection successes means that there are more calls that have reconnected after being forcibly terminated. On the other hand, a value less than unity means that there are many calls that have given up trying to achieve communications without making even one successful connection. The average number of connection successes is denoted as A_s and defined as follows:

$$A_s = N_s / C_n \tag{3}$$

Here, N_s is the number of connection successes and C_n is the total number of newly generated calls.

5 Simulation Results

The following presents the results of computer simulations for the evaluation measures described in Sect. 4 against the traffic load within the cells, the maximum retry number, and the speed of mobile station communicating in a cell. The symbols used in Figs. 2 to 7 are defined in Table 2.

Table 2. Difinition of symbols used in Figs.2 to 7

		Taking no account of reattempt call	Taking account of reattempt call
Conventional method	$C_h = 0$	--■--	--■--
	$C_h = 2$	--●--	--●--
	$C_h = 4$	--▲--	--▲--
Proposed method	$B_2 = 0.07$		--▽--
	$B_2 = 0.15$		--◇--

5.1 Call-Completed Rates

(a) Characteristics as Function of Traffic Load
When not taking reattempt calls into account, all calls that complete are calls that do so without being forcibly terminated during a call. In this case, the call-completed rate (R_{cl}) gradually drops as traffic load increases. In addition, for the same traffic load, R_{cl} becomes lower as the number of circuits reserved for handoff becomes larger. This is because more handoff circuits means less circuits that can be used for connecting new calls, that is, the number of calls that are blocked when attempting a connection increases.

When taking reattempt calls into account, the call-completed rate (R_{cl}) for only calls that do not experience a forced termination becomes higher as the number of circuits reserved for handoff becomes larger for the same traffic load, which is opposite the characteristics shown when not considering reattempt calls. Here, despite the fact that there are less circuits for connecting new calls, connection can be achieved by reconnecting, and the probability of forced termination becomes smaller once a connection is made. In addition, for small traffic load, R_{c1} shows a higher value than that when not considering reattempt calls, but for larger traffic load, it shows a lower value. This is because a larger traffic load means that circuits must carry more traffic for calls attempting reconnection than that when not considering reattempt calls, which in turn increases the probability of being blocked at handoff. As a characteristic of the proposed system, more handoff circuits are reserved as traffic volume increases in a cell. Next, the call-completed rate (R_{c2}) that includes reattempt calls is about unity up to a certain traffic load. In other words, most calls are eventually completed by reconnecting. The R_{c2}, however, begins to drop at a certain traffic load. This drop begins at smaller traffic loads as the number of handoff circuits increases. The proposed method exhibits similar characteristics with the case of reserving four circuits for handoff. For heavy offered traffic, R_{c1} is small for large values of C_h while R_{c2} is large, i.e., R_{c1} and R_{c2} have a tradeoff relationship. For the proposed system, if B_2 is set to a large value, R_{c1} decreases while R_{c2} increases. In this relationship between R_{c1} and R_{c2}, more importance can be attached to one or the other by adjusting the value of B_2. Call-completed rates versus traffic load are shown in Fig. 2.

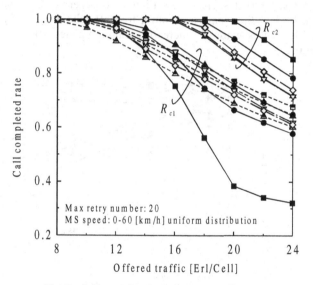

Fig. 2. Call-completed rates versus traffic load.

(b) Characteristics as Function of MS Speed

When not taking reattempt calls into account, the R_{c1} drops, if only slightly, as the MS picks up speed. The reason for this is that as speed increases, the frequency of handoffs likewise increases and the probability of being forcibly terminated becomes higher. For the same speed, the R_{c1} becomes lower as the number of circuits reserved for handoff increases.

When taking reattempt calls into account, the R_{c1} becomes lower as the number of handoff circuits increases at low speeds. As speed increases, however, R_{c1} is higher for more circuits reserved for handoff. Also, for a fixed number of handoff circuits, R_{c1} tends to increase with increase in speed up to a certain speed, but then decreases with further increase in speed. This can be attributed to the following. As handoff circuits increase, circuits that can be used for new calls decrease, and calls that give up connecting increase. Moreover, as MS speed is low, calls that have successfully connected will most probably stay inside the cell for the duration of the call. As speed increases, though, calls will soon move into adjacent cells. At this time, the probability of being blocked decreases as the number of handoff circuits increases. At slow speeds, the governing factor is block probability at connection time, whereas at high speeds, it is block probability at handoff time, resulting in the above characteristics. For fixed C_h, the R_{c2} is smaller for a larger number of C_h. Furthermore, for the same number of C_h, R_{c2} becomes slightly larger as the speed of the mobile station increases. In short, for fixed C_h, a larger value of C_h results in a smaller value for both R_{c1} and R_{c2} and thus degraded characteristics at low mobile station speeds. For higher mobile station speeds, a large set value of C_h results in a large value for R_{c1} and good characteristics but in a smaller value for R_{c2} and degraded characteristics. The proposed system sets C_h to a small value for low mobile station speeds and to a large value for higher speeds so as to prevent R_{c1} and R_{c2} from becoming small and degrading characteristics. Here, by setting C_h to larger values as mobile station speed increases, R_{c1} and R_{c2} again enter into a tradeoff relationship. If B_2 is set to a large value, R_{c1} decreases while R_{c2} increases. Call-completed rates versus MS speed are shown in Fig. 3.

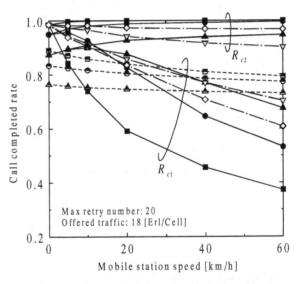

Fig. 3. Call-completed rates versus MS speed.

(c) Characteristics as Function of Maximum Retry Number

The call-completed rate (R_{c1}) for only calls that do not experience a forced termination becomes lower as the number of handoff circuits increases at small maximum retry number (N_r). As N_r increases, however, R_{c1} increases when more circuits are reserved for handoff. As more handoff circuits are reserved, circuits that can be used for new calls become smaller. Therefore, calls that give up connection increase, when N_r is small. Even if circuits that can be used for new calls are few, the call-completion probability increases, because the MSs can retry connection request repeatedly as N_r increases. After having succeeded in connection, if many handoff circuits are reserved, the probability of being forced termination become small. In the proposed method, though, the number of circuits reserved for handoff is controlled in accordance with N_r to reduce the number of forcibly terminated calls. In the conventional method, the call-completed rate (R_{c2}) that includes reattempt calls becomes lower as the number of handoff circuits increases. And R_{c2} increases with N_r when the number of handoff circuits is constant. In the proposed method, R_{c2} also increases with N_r, but slightly. If B_2 is set to a large value, R_{c1} decreases while R_{c2} increases. Call-completed rates versus maximum retry number are shown in Fig. 4.

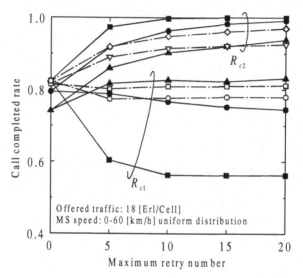

Fig. 4. Call-completed rates versus maximum retry number.

5.2 Average Number of Connection Successes

(a) Characteristics as Function of Traffic Load

When not taking reattempt calls into account, the A_s decreases slightly as traffic load increases. For the same traffic load, it decreases as the number of handoff circuits increases. This is because there is no attempt to reconnect after a failed connection in this case.

On the other hand, when taking reattempt calls into account, the A_s increases rapidly for a small number of handoff circuits up to a certain traffic load as traffic load increases. This is because, if there are only a few handoff circuits, there will be many calls that are forcibly terminated but then reestablish communications by reconnecting. Then, as traffic load further increases, the A_s will start to decrease. The reason for this is that as traffic load becomes excessive, there will be many calls that give up on establishing communications after attempting to reconnect time and time again without success. This decrease in A_s begins at a smaller traffic load as the number of handoff circuits increases. In addition, for many handoff circuits, the A_s drops below unity at high traffic load. This is interpreted to be due to the following. Since the number of circuits that can be used for connection is relatively small compared to traffic load, there will be many calls that cannot obtain a circuit when attempting to initiate communications for the first time and that then give up without connecting even once. In the proposed method, this average approaches unity regardless of traffic load within the cell. This is because the system controls the number of reserved handoff circuits to achieve a good balance between block at call attempt and forced terminations at handoff. Average number of connection successes (A_s) versus traffic load is shown in Fig.5.

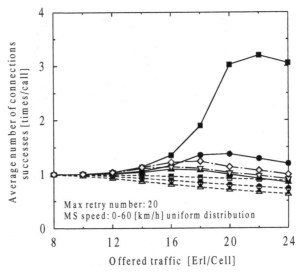

Fig. 5. Average number of connection successes (A_s) versus traffic load.

(b) Characteristics as Function of MS Speed
When not taking reattempt calls into account, this average is about constant regardless of the speed of the MS. On the other hand, when taking reattempt calls into account, the average increases as speed increases. This is because the probability of forced terminate becomes higher as speed increases and as handoff frequency increases, which in turn means that many calls will reconnect any number of times after being forcibly terminated. For the same MS speed, the A_s becomes smaller as handoff circuits increase. The reason here is that the number of circuits that can be used for connection becomes smaller as the number of handoff circuits becomes larger, resulting in many calls that give up on establishing communications after trying to reconnect many times without success. For slow speeds, the A_s drops below unity for many handoff circuits. In the proposed method, the A_s comes closer to unity

depending on the speed. The system controls the number of handoff circuits to achieve good balance between block at call attempt and forced terminations at handoff even if the average speed of MSs within the cell fluctuates. Average number of connection successes (A_s) versus MS speed is shown in Fig. 6.

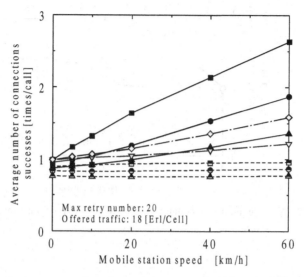

Fig. 6. Average number of connection successes (A_s) versus MS speed.

(c) Characteristics as Function of Maximum Retry Number

The A_s increases with maximum retry number (N_r), because the probability that the calls can reconnect after being forcibly terminated becomes higher as the N_r increases. For the same N_r, the A_s becomes smaller as the number of handoff circuits increase, because the number of circuits that can be used for initial connection becomes smaller as the number of handoff circuits becomes larger, resulting in many calls that give up establishing communications after trying to connect repeatedly without success. For small N_r, the A_s drops below unity. In the proposed method, the A_s comes closer to unity, slightly depending on the N_r. The system can control the number of handoff circuits to achieve good balance between block at call attempt and forced termination at handoff even if the N_r fluctuates. Average number of connection successes (A_s) versus maximum retry number is shown in Fig. 7.

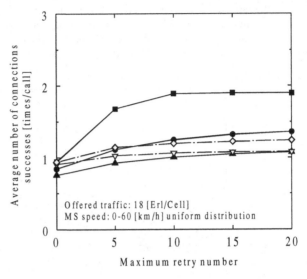

Fig. 7. Average number of connection successes (A_s) versus maximum retry number.

6 Conclusion

This paper has proposed a handoff priority control method that measures the forced termination rate at the time of handoff processing and that varies the number of handoff circuits according to the measured value. In addition, with regard to this method and one that reserves a fixed number of handoff circuits, we evaluated traffic characteristics when taking reattempt calls into account while comparing with the case that ignores reattempt calls. The following results were obtained. When considering reattempt calls, most calls eventually complete by reconnecting up to a certain traffic load. The system proposed here automatically adjusts C_h in response to an increase in handoff traffic resulting from an increase in offered traffic. This has the effect of suppressing a dramatic decrease in R_{c1}. When mobile station speed is low, C_h is set to a small value, which prevents degraded characteristics corresponding to small values of R_{c1} and R_{c2} brought on by an excessive number of C_h relative to speed. When mobile station speed is high, C_h is set accordingly to a large value, and R_{c1} and R_{c2} enter into a tradeoff relationship. More importance can be attached to either R_{c1} or R_{c2} at the time of system design by adjusting the value of B_2. The average number of connection successes increases rapidly as traffic load increases and handoff circuits decrease up to a certain traffic load, but decreases with further increase in traffic load. Moreover, if traffic load becomes great and the number of handoff circuits is high, the average number of connection successes drops below unity. Finally, it was shown that the proposed method makes it possible to mitigate the effects of change in average speed of mobile stations and in reattempt-call parameter such as maximum retry number.

References

[1] D. Hong and S. S. Rappaport, "Traffic model and performance analysis for cellular mobile radio telephone systems with prioritized and nonprioritized handoff procedures," IEEE Trans. Veh. Technol., vol. VT-35, no. 3, pp. 77-92, Aug. 1986.

[2] Qing-An Zeng, K. Mukumoto and A. Fukuda, "Performance Analysis of Mobile Cellular Radio System with Priority Reservation Handoff Procedures," in Proc. 44th IEEE VTC'94, pp. 1829-1833, June 1994.

[3] Chong Ho Yoon and Chong Kwan Un, "Performance of personal portable radio telephon systems with and without guarg chanels," IEEE J. Select. Areas Commun., vol. 11, no. 6, pp. 911-917, Aug. 1993.

[4] M. D. Kulavaratharasah and A. H. Aghvami, "Teletraffic performance evaluation of microcellular personal communication networks (PCS's) whit prioritized handoff procedures." IEEE Trans. Veh. Technol., vol. 48, no. 1, pp. 137-152, Jan. 1999.

[5] S. Tekinay and B. Jabbari, "A measurement-based prioritization scheme for handovers in mobile cellular networks," IEEE J. Select. Areas Commun., vol. 10, no. 8, pp. 1343-1350, Oct. 1992.

Threshold-Based Registration (TBR) in Mobile IPv6

Linfeng Yang, Jouni Karvo, Teemu Tynjälä, and Hannu Kari

Helsinki University of Technology, Telecommunications Software and Multimedia
Laboratory, P.O.Box 9700, 02015 HUT, Finland
{lyang, kex, tjtynjal, hhk}@tcm.hut.fi
http://www.tcm.hut.fi/english.html

Abstract. The underlying principles of IETF Internet Draft, Mobility
Support in IPv6, make it possible to employ some mechanisms to improve
handoff smoothness, to maintain optimized data transfer route, at the
mean time without requiring any special support from the network side.
Such mechanisms were introduced and examined in this paper together
with the proposal of Threshold-Based Registration in Mobile IPv6. In
this proposal, a mobile node will establish immediate forwarding from
the previous care-of address whenever it moves to another subnetwork.
After every fixed number of immediate forwarding steps, the mobile node
will establish direct forwarding from its primary care-of address. Again,
after every fixed number of direct forwarding steps, the mobile node will
register a new primary care-of address to its home agent. With such an
approach, the above mentioned goals are achieved.

1   Introduction

Internet Protocol version 6 (IPv6)[1], is currently under development. As an
evolution of current Internet Protocol (IPv4), it provides many advantages, the
most important one being the support of mobility. Internet draft *Mobility Support in IPv6*[2], also known as MobileIPv6, proposed by the Internet Engineering
Task Force (IETF), gives strong support for route optimization compared with
its counterpart in Mobile IPv4[3].

In MobileIPv6, not only can the routes between mobile nodes and correspondent nodes be optimized, but also the route between a mobile node and its home
agent. As a disadvantage of this optimized route, high signaling load is incurred.
A proposal using a hierarchical approach to reduce this kind of signaling load has
been introduced[4]. However, this approach has been criticized for contradicting
the design goals of MobileIPv6[5].

The newest version of MobileIPv6 has clarifed how to establish forwarding
from the previous care-of address. It intends to reduce packet loss during handoff.
We claim that this mechanism can also be employed to reduce the signaling
load. This article will present a scheme for using forwarding to reduce signaling
load while maintaining optimized data routes between correspondent nodes and
mobile nodes.

C.G. Omidyar (Ed.): MWCN 2000, LNCS 1818, pp. 150–157, 2000.

The paper is organized as follows: Section 2 describes the current method for mobility support in IPv6. Section 3 presents our threshold based proposal. Section 4 contains a brief discussion on the proposed approach, and Section 5 concludes the paper.

2 Mobility Support in IPv6

In this section, we describe the current method for mobility support in MobileIPv6[2]. First, some terminology needs to be defined. A *home link* is the link by which a mobile node's home subnet prefix is defined. Standard IP routing mechanism will deliver packets destined for a mobile node's home address to its home link. A *care-of address* means an IP address associated with a mobile node when the mobile node is visiting a foreign link; the subnet prefix of this IP address is a foreign subnet prefix. A router on a mobile node's home link with which the mobile node has registered its current care-of address is called a *home agent*. The care-of address registered with the mobile node's home agent is called the *primary care-of address*.

A *correspondent node* is a node that wants to send a packet or several packets to the mobile node. A *binding* is the association of the home address of a mobile node with a care-of address for that mobile node, along with the remaining lifetime of that association. Each IPv6 node has a *binding cache*, which contains binding related information. A *binding update* is an IPv6 Destination Option used by a mobile node to notify a correspondent node or the mobile node's home agent of its current binding. That is, it can be piggybacked on data packets of the ongoing connection, and only the destination node can process this option. A *routing header* is an IPv6 option header; it is used by an IPv6 source to list one or more intermediate nodes to be "visited" on the way to a packet's destination[1].

When a correspondent node wants to send packets to a mobile node, it first examines its binding cache for an entry for the destination address to which the packet is being sent. If no entry is found, it simply sends the packets normally, with no routing header. If the mobile node has not registered a care-of address with any home agent, the packets will go to the mobile node's home link as conventional IPv6. If the mobile node is currently at a foreign network and had registered a care-of address to this home agent, the packet will be received by the mobile node's home agent. The home agent then encapsulates the packet and tunnels it to the mobile node's care-of address. Upon receiving the tunneled packet, the mobile node will send a binding update to the correspondent node. The correspondent node can then send the rest of the packets directly to the mobile node using a routing header. See Fig. 1.

MobileIPv6[2] Sec. 10.5, Forming New Care-of Addresses says:

> "After detecting that it has moved from one link to another (i.e., its current default router has become unreachable and it has discovered a new default router), a mobile node SHOULD form a new primary care-of address using one of the on-link subnet prefixes advertised by the new

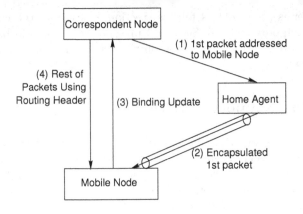

Fig. 1. Routing in MobileIPv6

router. A mobile node MAY form a new primary care-of address at any time, except that it MUST NOT do so too frequently."[1]

And Sec. 10.6, Sending Binding Updates to the Home Agent says:

"After deciding to change its primary care-of address as described in Sections 10.4 and 10.5, a mobile node MUST register this care-of address with its home agent in order to make this its primary care-of address."

So, if we can form a care-of address and use it to maintain mobile node accessibility, without registering it to the home agent, then the signaling load will be reduced. MobileIPv6 also permits a mobile node to form a new primary care-of address at any time, so we may postpone forming a new primary care-of address when it is possible. Our approach is based on this observation, and will be described in the next section.

3 Threshold-Based Registration (TBR)

To maintain mobile node accessibility without forming a new primary care-of address each time a mobile node moves to another subnetwork, we employ the mechanism introduced in MobileIPv6 Section 10.9. Establishing Forwarding from a Previous Care-of Address.

First, we define a new terminology, *anchor home agent*. When a mobile node moved to a foreign link and registered a primary care-of address to its home agent, the router in this foreign link which can act as the home agent of this mobile node's primary care-of address will be called an anchor home agent. Later,

---

[1] The keywords "SHOULD", "MUST", and "MAY" are frequently used in RFCs. They are required by MobileIPv6 to be interpreted as described in *Key words for use in RFCs to indicate requirement levels, RFC 2119*[6].

the mobile node can establish *direct forwarding* from this anchor home agent to its current care-of address. An *immediate forwarding* means that packets are forwarded from a mobile node's previous care-of address to the current care-of address as described in MobileIPv6 Sec. 10.9.

In the following subsection, we will introduce employment of this forwarding mechanism by giving three approaches. The third and final proposal of TBR is indeed the compromise of the first two.

3.1 First Approach

In the first approach a mobile node establishes forwarding only from its immediate previous care-of address. It repeats such forwarding until either there is no router in the previous visited subnet that can act as a home agent or the number of forwarding steps exceeds a predefined limit. This limit is called the maximum number of forwarding steps F_{max}. At this time, the mobile node will form a new primary care-of address using the newest care-of address and sends a binding update to its home agent. See Fig. 2 for a possible scenario.

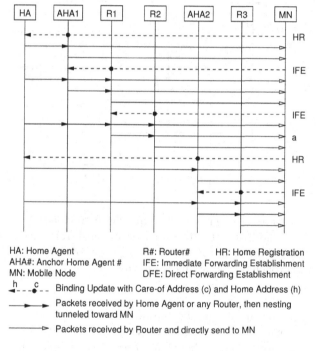

HA: Home Agent   R#: Router#   HR: Home Registration
AHA#: Anchor Home Agent #   IFE: Immediate Forwarding Establishment
MN: Mobile Node   DFE: Direct Forwarding Establishment

$\overset{h}{\leftarrow}$ - - $\overset{c}{\bullet}$ - - Binding Update with Care-of Address (c) and Home Address (h)

⟶ Packets received by Home Agent or any Router, then nesting tunneled toward MN

⟶ Packets received by Router and directly send to MN

Fig. 2. Establishes forwarding from previous care-of address only. Notice the start point of received packets at *a*, only those close to mobile node's current location are shown in the figure. This is due to direct binding update to the correspondent node, which then sends packets to mobile node's most recent care-of addresses

In practice, this approach introduced a nested tunnel which may lead to segmentation of data packets and add delay for data processing. *Generic Packet Tunneling in IPv6 Specification*[7] recommends that the default value of IPv6 Tunnel Encapsulation Limit is 4. If a packet with Destination Option extension header containing this limit option with value close to 4 is sent by a correspondent node, and the nested tunnel contains more than this number of forwarding, the packets will be discarded before reaching the mobile node. Thus, the benefit of this approach is limited.

3.2 Second Approach

Another approach would be that the mobile node establishes forwarding both from the immediate previous care-of address and from its current primary care-of address when it moves to another subnetwork. See Fig. 3.

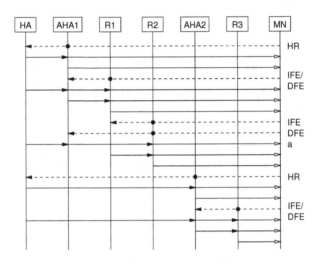

Fig. 3. Establishes forwarding from both immediate previous care-of address and the primary care-of address. At *a*, the packets tunneled from the home agent will need only one extra tunnel to the mobile node, thus solving the problem arising from the first approach. Other symbols are the same as in Fig. 2

Note that in this approach, the lifetime used in the binding updates for establishing forwarding from immediate previous care-of address can be set short. There is no danger of breaking the chain of forwarding, since there is always a direct route from the anchor home agent to the mobile node. But this approach generates a double signaling load locally, and it also consumes more of the mobile node's resources than the first approach.

3.3   TBR — Final Proposal

Thus, we propose a final approach, the TBR approach. In this approach, a mobile node establishes immediate forwarding from the previous care-of address each time it moves to another subnetwork. After a small number of successful immediate forwarding steps, denoted by IF_{max}, the mobile node will establish a direct forwarding from its current primary care-of address. After a number of successful direct forwarding steps, denoted by DF_{max}, the mobile node will register to its home agent using the latest care-of address as a new primary care-of address. The mobile node will register the latest care-of address to its home agent also whenever the remaining lifetime of the last home registration is near to the expiry, as recommended by MobileIPv6. See Fig. 4 as an example.

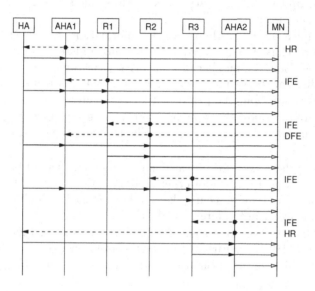

Fig. 4. Mobile node establishes forwarding with TBR approach with $IF_{max} = 2$ and $DF_{max} = 2$. Symbols are the same as in Fig. 2 and Fig. 3

3.4   Tunneling Loop Avoidance

In order to avoid packets tunneled in a loop between two or several routers, a mobile node should act as returning home described in MobileIPv6 Section 10.17 when it switches to a subnetwork from which the forwarding is still valid. It should send a binding update to this router with a zero lifetime, to de-register the previous binding before establishing forwarding from the immediate previous care-of address.

4   Discussion

The benefits of the TBR approach are as follows:

1. Achieving balance between optimized routes and low signaling load.
2. Scalability and reliability. Since a mobile node will change its anchor home agent whenever the lifetime for home registration or the limit number of direct forwarding steps DF_{max} is reached, the load for the anchor home agents is distributed among the routers in the foreign network. When the number of mobile nodes in the foreign networks increases, there is a smaller probability that the anchor default router will become a bottleneck. When one of the routers is out of order, the mobile node can change its anchor home agent by updating its primary care-of address and sending a new registration to the home agent.
3. Easy implementation. Almost all the information we needed to implement this approach is already stored in the binding update list of the mobile node. Only two counters (number of immediate forwarding and number of direct forwarding) are needed to make a quick decision on which kind of forwarding or registration should be performed for each handoff.
4. Transparent support. Each mobile node can implement this approach without pre-negotiation between the mobile node and the foreign network.

Binding updates used for forwarding establishment should be authenticated in the same way as described for other binding updates in the MobileIPv6. This introduces a need for more security keys to be managed by both the mobile node and the home agent in the foreign network.

Although, we have done some reachability analysis of the approach with formal methods[8], but clearly a thorough performance analysis left for future work, in order to find the optimal values for the parameters IF_{max} and DF_{max}.

5   Conclusion

In this article, we presented the current state-of-the-art MobileIPv6, concentrating on routing and packet forwarding. We presented a modified algorithm for binding updates for reducing signaling load while maintaining data route optimality. The algorithm is, a mobile node establishes immediate forwarding from the previous care-of address whenever it moves to another subnetwork. After every fixed number of immediate forwarding steps, the mobile node establishes direct forwarding from its primary care-of address. Again, after every fixed number of direct forwarding steps, the mobile node registeres a new primary care-of address to its home agent. With such an approach, the above mentioned goals are achieved together with scalabilty, reliability, easy implementation and transparent support as the benefits.

References

1. Deering, S., Hinden, R.: Internet Protocol version 6 (IPv6) specification. RFC 2460. 1998
2. Johnson, D., Perkins, C.: Mobility Support in IPv6. Internet Draft. Work in progress. 2000. http://www.ietf.org/internet-drafts/draft-ietf-mobileip-ipv6-10.txt
3. Perkins, C., Editor: IP Mobility Support. RFC 2002. 1996
4. Castelluccia, C., Bellier, L.: Toward a Unified Hierarchical Mobility Management Framework. Internet Draft. Work in progress. 1999. http://www.ietf.org/internet-draft-castelluccia-uhmm-framework-00.txt
5. IETF Online Proceedings, Forty-fifth IETF, Oslo, July 11 - July 16, 1999, Work group 2.5.2 IP Routing for Wireless/Mobile Hosts (mobileip)
6. Bradner, S.: Key words for use in RFCs to indicate requirement levels. RFC 2119. 1997
7. Conta, A., Deering, S.: Generic Packet Tunneling in IPv6 Specification. RFC 2473. 1998. 20
8. Tynjälä, T., Kari, H., Yang, L.: Verification of Threshold-Based Registraion Algorithm in Mobile IPv6. Submitted to IEEE International Conference on Networks' 2000 (September 2000). 2000

Enhanced Mobile IP Protocol

Baher Esmat, Mikhail N. Mikhail, Amr El Kadi

Department of Computer Science, The American University in Cairo,
Cairo, Egypt
{besmat, mikhail, elkadi}@aucegypt.edu

Abstract. One of the most recent Internet challenges is to support transparent movement of people along with their computers, data and most of all applications. Therefore, Mobile IP has been developed to provide Internet mobility services.

This paper aims at enhancing the IETF Mobile IP standard. The model developed in this paper suggests a new caching mechanism, which is based on the Mobile Information Server (MIS). Actually, the MIS is designed to be part of the border router of any network that supports mobility services. Moreover, the paper suggests a *peering technique* by which information about mobiles hosts could be shared among different MISs. All the design issues including model components as well as mechanisms for caching and peering are described in details.

The simulation results show that the proposed design provides improved performance and better bandwidth utilization. The suggested architecture provides other qualitative advantages such as scalability and transparency.

1 Introduction

Mobile computing has assumed an increasing importance in recent years, and will pervade future distributed computing system. Although network standards were not designed with the capability of supporting the demand of mobility, the need is that they should grant the users a continuous access to their data, irrespective of their point of attachment. Mobile computing is still restricted by many obstacles [1].

As a mater of fact, the current IP version 4 [2] makes an implicit assumption that the point at which a computer is attached to the Internet is fixed, and its IP address identifies the network to which it belongs. The challenge is to develop a protocol, which allows computers to roam freely around the Internet and communicate with other stationary or mobile nodes, without major changes in the existing TCP/IP stack.

The mobility problem within the Internet is mainly concerned with the IP layer, since this layer handles all aspects related to addressing as well as routing. To illustrate this point [3], if a computer moves to another network, and retains its original IP address, this address will not reflect its new location, and consequently, all routed packets to this host will be lost. In the other hand if the mobile host gets a new address when migrating to another network, the IP address changes, the transport layer (i.e. TCP) connection identifier changes too [4], and hence all connections with this mobile host through its old address are going to be lost. Therefore, if the mobile

C.G. Omidyar (Ed.): MWCN 2000, LNCS 1818, pp. 158-173, 2000.
© Springer-Verlag Berlin Heidelberg 2000

host moves without changing its address, it will lose routing, and if it gets new address, it will lose connections.

This paper is organized as follows. The next section presents the Mobile IP standard protocol. Section 3 describes the contribution of this work. An overview for the proposed design is going to be illustrated in section 4. In section 5, all the model components will be identified and discussed in detail. Next, section 6 presents all the simulation details as well as the results. Finally, section 7 concludes the paper.

2 Mobile IP

During the last few years, many contributions have been offered by different entities and groups, towards designing a model for a mobility supports Internet. The proposed models are different in terms of their components and methodology, but they are all aiming at keeping the mobile hosts communicating transparently via the Internet. Proposals from Columbia University [5,6], Sony [7,8], the Loose Source Routing(LSR) Proposal [9] as well as the Internet Engineering Task Force (IETF) Mobile IP working group [10,11,12], are the most outstanding models,

Since this work is an extension to the IETF Mobile IP standard, it is worth to focus on the operation of this standard protocol. First of all, it should be mentioned that the Mobile IP working group has been in charge of standardizing Mobile IP. Recently, the Mobile IP has become a standard, after passing through two stages [13]. The first one in which the base protocol was developed, with the objective that mobile nodes can roam transparently around the Internet, with no modifications whatsoever to other stationary nodes. The second phase has answered many open questions regarding the best route that the packet may take to reach a mobile node. This has been known by the *route optimization* problem.

2.1 Mobile IP Operation

According to [12], the IETF Mobile IP architecture defines special entities called the Home Agent (HA) and the Foreign Agent (FA), both cooperate to allow a Mobile Host (MH) to move without changing its IP address. Each MH is associated with a unique *home network* as indicated by its permanent IP address. Normal IP routing always delivers packets meant for the MH to this network. When an MH moves to a *foreign network*, the HA is responsible for intercepting and forwarding packets destined to the MH to anew address which is called the *care-of address*. The MH uses a special registration protocol to keep its HA informed with its new location.

Whenever a MH moves from its home network to a foreign network, or from one foreign network to another, it looks for a FA on the new network in order to obtain its new care-of address. In order for the MH to be able to work with this new address, it must go through a registration procedure via both, the foreign agent and the home agent. After a successful registration, packets arriving for the MH on its home network are *encapsulated* by its HA and forwarded to its FA. Encapsulation refers to the process of enclosing the original datagram as data inside another datagram with new IP header [14]. The source and destination addresses in the new header correspond to the HA and FA respectively. Upon receiving the encapsulated datagram, the FA strips off the new header and forwards the original one to the MH. This process at the FA end is known as *decapsulation*. If on the other hand, the mobile node needs to send a packet to any destination, the packet will be routed to its destination with the normal fashion without using either the home agent or the foreign agent. The figure below illustrates the operation of the mobile IP routing.

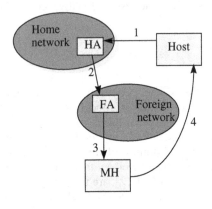

Fig. 1.

2.2 Route Optimization

As depicted in Figure 1, the IETF scheme has a major routing problem. Any packet which means to reach a mobile host, is directed first to the appropriate home agent, then to the foreign agent and finally received by the mobile host. This means that the above scheme does not allow the source host to reach the mobile host directly without passing by its home network. This problem is known as the *triangle routing problem*, and in order to solve it, a technique for route optimization is needed.

Route optimization [15] means solvingthe problem of triangle routing, by allowing for each host to maintain a binding cache for a mobile host wherever it is. When sending a packet to a mobile node, if the sender has a binding cache containing the care-of address of that mobile node, it will deliver the packet directly toward the mobile node, without the need to pass through the home network.

2.3 Mobile IP Problems

Although the IETF Mobile IP working group has enhanced its base protocol and provided a solution for route optimization, the protocol seems to have some deficiencies. From a performance point of view, the protocol intended to optimize the routing process, but realistically the routing has not been thoroughly optimized. For example, consider two computers connected to the same network, computer A and computer B. If the first one wants to reach a certain mobile host, it must go through the home network, at least for the first few packets, before reaching the mobile node.

Then, it caches the mobile node's address, so that it could reach it directly for the rest of the packets. Afterwards, if computer B needs to access the same mobile host, it must go through the same procedure again, as it does not have any cached information regarding the mobile host. The same process applies for any host on this network when trying to contact this specific mobile host. This implies that the first few packets directed from any host on a certain network toward a mobile host, are inefficiently routed through the home agent.

Another problem exist that relate to the fact that current implementation of Internet Protocol version 4 (IPv4) [2], which is currently running on all the Internet hosts worldwide, do not allow for any such mobile information to be cached. This means that in order for such a routing optimization to be achieved, every single host on the Internet must have its IP software modified.

2.4 Mobility Support in IPv6

IPv6 [16] has been developed with some sophisticated features that have not been supported by the current Internet Protocol (IPv4). IPv6 sustains major requirements concerning addressing, routing, security and mobility.

Mobility support in IPv6 [17] follows the same methodology that has been developed for IPv4. The same terminology is still valid as for home agents, mobile hosts, home and foreign networks, as well as encapsulation or tunneling. However, the term foreign agent is not of any more use. The reason is that any IPv6 is able to configure its own IP address automatically, as well as to choose its default gateway. This is accomplished via the Stateless Address Autoconfiguration [18] and the Neighbor Discovery protocols [19]. Therefore, it is quite straightforward that whenever an IPv6 mobile host migrates to any foreign network, it could easily detect the change in network connectivity, and configure its IP address automatically. Moreover, mobility within IPv6 borrows heavily from the route optimization specified for IPv4, which was described earlier in a previous section. By default, all IPv6 hosts are able to cache mobility information, via authenticated binding update messages. It is only the mobile host that has the authority to send binding updates to any other correspondent nodes.

Although people have been waiting for IPv6 to become the Internet standard [20], and many vendors have implemented IPv6 in their products for testing purposes, IPv6 is still under development. It is quite conceivable that the Mobile IP deployment will coincide with the standardization and implementation of IPv6 [21].

3 The Proposed Scheme

This work is intended to enhance the Mobile IP standard that has been developed by the IETF Mobile IP working group. Most of the Mobile IP protocol specifications are used in the development of this work.

This paper suggests a method for caching mobile information, different from that developed by the IETF working group. The proposed model implies suggests a *central cache engine* within each network, or a cluster of networks, responsible for caching mobile information. Moreover, all the functions performed by the HA's and

FA's, regarding encapsulation, decapsulation, registration and authentication, could be part of this central cache engine.

In addition, it is recommended for any network, or class of networks, connected to the Internet, to use its boarder router as a Mobile Information Server (MIS) which handles all caching as well as mobility services. As a matter of fact, designing a central caching mechanism does not necessarily imply that this should be part of the network router. Instead, building such a cache server, along with other mobility functions, can take place in any workstation in the network. However, this design is recommended for more than one reason. First, the proposed model manipulates the cached mobile information as part of the routing information that already exists in the routers, so that any cache entries are considered part of the routing table. . This new model allows for MISs to work in a peering fashion, by which mobility information can be exchanged. In addition, the model developed here aims at being transparent for the IP version used, whether it is IPv4 or IPv6. Eventually, the paper delivers a new caching mechanism, as part of the Mobile IP protocol. A complete practical architecture, with a simulation of all components and their functionality is delivered. Efficiency, scalability and transparency are the main value-added features in this new scheme, taking into account all security policies which have been addressed through the base Mobile IP model.

4 Design Overview

The proposed design is based on a centralized caching architecture. For a specific network, there is a cache server responsible for any mobile information concerning any node belonging to that network, or even it could cache other information regarding any external mobile node. In addition to caching, this server can handle all the functions of the home agent as well as the foreign agent, such as registration, authentication and tunneling procedures.

5 Model Components and Description

Figure 2 depicts the main components of the new suggested design. The figure illustrates four different networks (any networks that are members o the Internet for demonstration purpose) in order to describe the various functions and scenarios of this model.

5.1 Mobile Information Server (MIS)

The new model defines a new term called MIS. The MIS is suggested to be implemented in the border router. Border routers are basically responsible for routing the traffic between a group of networks and the outside world of the Internet. Moreover, In addition, border routers are now made responsible for other mobile services that were part of the home agent and the foreign agent in the Mobile IP scheme. Also, the new caching mechanism is designed to take place on these routers.

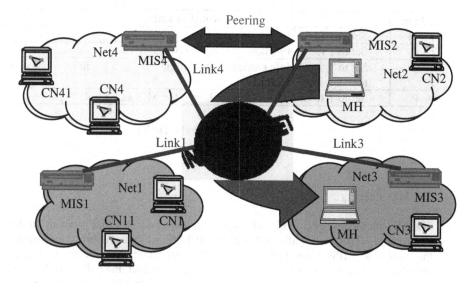

Fig. 2. New proposed model

Therefore, and because any of these routers handles a lot of other new services, it is going to be called in the context of this paper the MIS.

In addition to routing and caching functions, any MIS can be configured to work as a peer to another MIS. As a result of peering, MISs can exchange mobile information by the same concept of exchanging routing updates. Moreover, the MIS should include a table that encompasses all IP addresses of visiting mobile hosts, along with their Media Access Control (MAC) addresses, in order to deliver the packets to the proper destination after decapsulation. This table is called the *visitor list*.

5.2 Caching

It has been mentioned earlier that the MISs are responsible for caching mobile information. Actually, the cache entries are considered normal routing entries, with some extra fields for mobility.

Basically, the cache entry is suggested to include the following fields:

M O C P   *old IP address*   *next hop*   *new IP address*

In this entry, the first four fields are flags which indicate the type of this routing entry. The following table describes briefly each of those flags.

The next two fields are similar to any ordinary routing information, they represent a destination IP address, as well as the next router in the way to reach this destination. In this case, the destination IP address is the old IP address of the mobile host. Finally, there is another field that indicates the new location of the mobile node.

Flag	Routing Entry
M	Set to 1 for all mobile routing entries
O	Set to 1 for mobile hosts controlled by the MIS
C	Set to 1 for mobile entries cached at the MIS as a result of cache update messages
P	Set to 1 for mobile entries cached at the MIS as a result of peering

Table 1. Mobile routing entry

5.3 MIS Peering

The MIS methodology provides another value-added service that is not in the Mobile IP protocol. This model allows for different MISs to share mobility information. In this case, the MISs seem to be working as peers, and the process of exchanging mobility information is called peering. As a matter of fact, the MIS in order to export any information regarding certain mobile host to another MIS, it should be the owner of this mobile host. In other words, the routing entry for this specific mobile node must have its *O* flag set to 1. When this routing entry gets to the peer MIS, it will have the *P* flag set to 1, in order to indicate that this information has come as a result of peering. Moreover, each MIS has in its Mobile IP configuration a list in which other MIS peers are defined. This means that MISs work as peers to each other based on some pre-defined routing configuration.

5.4   Mobility Scenarios

Reference to Figure 2, this section describes a scenario in which a mobile host *MH* migrates from its home network *Net2* to a foreign network *Net3*. The scenario shows how other correspondent nodes can reach *MH* while it is away from *Net2*, and even when it gets back.

1. *MH* migrates to *Net3* and gets a care-of address, which is actually the IP address of *MIS3*. Then, it sends a registration request to *MIS3*, which will relay this request to *MIS2*.
2. *MIS2* accepts the registration after checking the authentication extension included in the request. Hence, a registration reply is sent to *MIS3*, which in turn will inform *MH* with its status.
3. From now on and since the registration request has been accepted, *MH* becomes reachable via *MIS3*.
4. *CN1* wants to reach *MH*. So, it will start sending packets toward *Net2*. The packets arrive at *MIS2*, which could realize that they are destined to *MH*. *MIS2* contains a cache entry for the new location of *MH*. Actually, the *O* flag for this cache entry must be set to 1 because *MH* is owned by *MIS2*.
5. Right after the encapsulation, *MIS2* recognizes that *CN1* does not perceive that *MH* had moved. Therefore, *MIS2* sends *CN1* a cache update message. This cache update will be intercepted by *MIS1*, which in turn will update its routing table. In

this case, the cache entry at *MIS1* will have its *C* flag set to 1 since it has been generated through a cache update message.

6. *MIS3* receives encapsulated packets, and decapsulates them to *MH*, after obtaining its MAC address from the visitor list.

7. All packets from *MH* in their way back to *CN1*, are always routed directly to *Net1*, without necessarily passing by *Net2*.

8. *MH* decides to return to *Net2*. Consequently, it will ask for registration at *MIS2*.

9. If *MIS2* accepts this registration request, *MH* starts acting normally without any The next two fields are similar to any ordinary routing information, they represent a destination IP address, as well as the next router in the way to reach this destination. In this case, the destination IP address is the old IP address of the mobile host. Finally, there is another field that indicates the new location of the mobile node. mobility services.

10. *MIS2* sends a cache delete message to *MIS3* in order to release any routing information regarding *MH*.

11. *CN1* may try again to contact *MH*.

12. Packets are going to be tunneled through *MIS1* and directed to *MIS3*. From decapsulation, *MIS3* will discover that the destination address is not any more in its mobile visitor list.

13. *MIS3* sends a cache delete message to *MIS1*, which consequently is going to route the packets without any kind of encapsulation, through the ordinary route to *MIS2*.

In addition, Figure 2 depicts another correspondent node *CN11* that may need to access *MH* *net3*. Unlike the standard Mobile IP, all the packets from *CN11* toward *MH* will be routed directly to *MIS3*, since *MIS1* has a cache entry for *MH*.

Another difference between this new architecture and the Mobile IP one is the peering mechanism. Again from Figure 2, *MIS2* and *MIS4* work as peers to each other. Hence, *MIS4* will be notified that *MH* had moved. Therefore, it is possible that *CN4* can contact *MH* directly through a tunnel from *MIS4* to *MIS3*.

6 Simulation

Throughout the development of this work, simulation has been used to compare the Mobile IP standard protocol and the new proposed model. The main prominent difference between the two models is the caching methodology, by which the route optimization is verified, and the whole routing performance is induced. Actually, the influence of the caching mechanism can be evident in the total amount of traffic through the whole network, as well as the delay associated with packets while carried from source to destination nodes.

Therefore, this simulation has focused on the traffic as a main point for discrimination. All the quantitative results shown out of the two comparable models are in terms of packet delay as well as bandwidth consumption.

6.1 COMNET III

COMNET III [22] is a performance analysis tool which simulates computer and communication networks. It can be used to model both circuit switching and packet switching networks. In addition, it can accommodate different topologies of WANs and LANs in which many standards and protocols are supported, such as Ethernet, Token Ring, PPP, X.25, Frame Relay and ATM.

Regarding this work, COMNET III version 1.2 for Windows has been used to implement a number of models which differentiate between the standard Mobile IP architecture proposed by the IETF, and the model suggested in this paper. The simulated model presents all the issues described previously, concerning registration, tunneling, caching, route optimization as well as border routers which are known here by MISs. The networks described in the simulation were represented by Ethernet connections for LANs, Point-to-Point (PPP) links for WANs and processing nodes to simulate computers and workstations. All the networks are inter-connected using routers, in which user-defined routing tables are used to simulate the model. In addition, sending and receiving messages among the various nodes simulates the traffic. Finally, the model is verified and executed, and the results can be shown in graphs, or presented through reports of text format.

6.2 Network Components

This section provides a description for the network components simulated by COMNET III.

Processing node: All the simulated models use the processing node component to describe generic Internet hosts which are considered the endpoints of any Internet traffic.

Router node: Routers have been configured with *500 Mbps* bus rate, *50000 packet per second* as a processing rate, and the input and output delays are ignored. Moreover, all the models developed in this simulation have standardized on the static routing protocol, since using any dynamic protocols will make no difference to the results.

LAN connectivity: The IEEE 802.3 Ethernet standard is used.
WAN link: Point-to-Point Protocol (PPP) is used for all communication links with a bandwidth of *1.536 Mbps.*

Message source: Message sources are used to represent specific traffic based on the TCP/IP Protocol. A payload of *1460 bytes* and header of *40 bytes* is used for all the messages generated throughout the simulation. In addition, the message size may be changed based on the type of the message itself.

6.3 Simulated Models

This section describes a number of network architectures that have been simulated throughout this work. Generally, all the simulated models present the differences between the Mobile IP standard, and the new proposed scheme. The simulated models reflect the scenario that has been previously illustrated in Figure 2, in which a mobile host is migrating from one network to another whilst a correspondent node is trying to reach it. The models are simulated in simple as well as complicated structures. Simple models aim at presenting a preliminary overview for the Mobile IP operation, focusing on the main mechanisms and services for each architecture. On the other side, other more advanced designs are required in an attempt to simulate something close to reality. Such composite models include many nodes, routers and message sources that load the network with much more traffic.

6.3.1 Simple Models

This section illustrates the simplest cases for any Mobile IP architecture, where the simulated models consist of a single mobile node that migrates from its home network to another foreign one. Besides, there is a correspondent node belonging to another third network and it wants to get access to the mobile node. This scenario is shown for both the Mobile IP standard with route optimization support, and the new architecture developed within this paper. The most outstanding difference between the two schemes is the caching mechanism, as well as the fact that the border router within the new model is responsible for all mobile services.

Moreover, the simulation shows all the procedures defined by the Mobile IP standard. Such procedures include the registration request messages, registration acceptance, encapsulation as well as decapsulation. According to [3], the size of the registration message is *24 bytes* plus variable length extensions required for authentication. Likewise, the registration-reply is *16 bytes* beside those needed for authentication. As per our simulation, the registration messages are *48 bytes*, whereas the registration-reply messages are *40 bytes*, since extra bytes are used to indicate the variable length authentication extensions.

6.3.2   Composite Models

Similar to the simple models, in which the new proposed design has been compared to the Mobile IP standard, the composite models perform the same analogy accompanied by adding more components to the simulated models. In addition, the generated traffic is much more than that generated for the simple models.

Actually, the composite models contain five networks, two mobile hosts as well as many other correspondent nodes. Moreover, the new model simulates the peeing methodology that has been developed throughout this work.

6.4   Results

This section presents all the results that have been collected as an output from executing the simulated models. It will be noticed that all the models have been simulated over a simulation time of *60 seconds*.

As per the simple models, the point of discrimination between the two simulated schemes is the delay for the packets running from the source network to the foreign network where the mobile node is located. On the other hand, the evaluation of the composite models is based more the bandwidth consumption.

6.4.1   Simulation Results for Simple Models

As for the Mobile IP route optimization standard and as illustrated in Figure 2, when *CN1* talks to *MH*, the first few packets are going to be routed via *Net2*, then encapsulated toward *Net3*. The delay of the *CN1-Msg1* packets as well as that of the encapsulated packets *HA-Encap* are illustrated in Figure 3(a) and 3(b) respectively. The total average delay is *161.018 msec*.

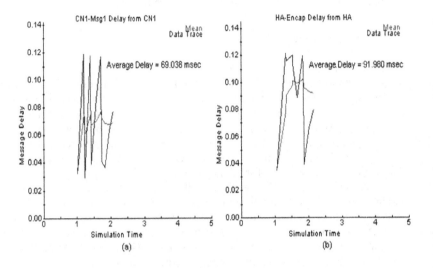

Fig. 3. Packet delay for *CN1-Msg1* and *HA-Encap*

But afterwards, *CN1* should get its cache updated and therefore packets are going to reach *MH* directly (CN1-Msg2) with an average packet delay of *94.544 msec*, as shown in Figure 4. The same scenario is applied for *CN11* when trying to access *MH*, causing almost the same results.

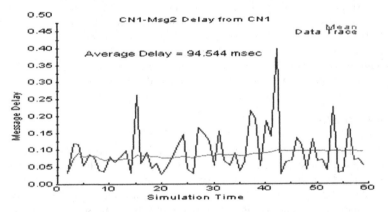

Fig. 4. Packet delay for *CN1-Msg1*

As formerly shown in Figure 2, *MIS1* is considered a central cache engine for *Net1*, responsible for any mobile information. Unlike the route optimization model, *CN11* may reach *MH* directly since the *MH* new care-of address is cached within *MIS1*. Therefore, the overall delay will be less than that of the last illustrated model. Figure 5 shows that in the new model, the average packet delay from *CN11* to *MH* is *79.681 msec*.

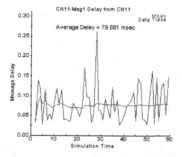

Fig. 5. Delay from *CN11* to *MH* in the new model

Although the difference in packet delays may be significant in such simple architectures, this might not be the case in real networks that carry millions of transferred packets per second. However, those models were basically simulated in order to prove that the proposed model has a quantitative advantage over the other models, even if this advantage is a minor issue in real applications. More importantly, this difference could be more significant from another perspective. For instance, if there is a certain application which ought the mobile node to stay in contact with a number of remote nodes at some other network, so that there are many nodes like *CN1* and *CN11* belong to the same network, and try to reach the same mobile host. In this case, it is better from a scalability point of view to have a central caching rather than storing the same information on many different machines.

6.4.2   Simulation Results for Composite Models

It has been stated before that for the composite models, the evaluation criteria is according to the bandwidth utilization. In fact, both composite models have been simulated with two networks working as home networks for two different mobile hosts, and each network has a single link to the Internet. The bandwidth utilization of each link is illustrated in this section.

Before presenting the results, it should be mentioned that COMNET III deals with any communication connection as a full-duplex link, in which the input bandwidth is independent of the output bandwidth. Therefore, it will be noticed that each link is represented by two graphs (a) and (b).

Assuming that the Internet link for the first home network is *Link A*, and for the other network is *Link B*. Figures 6 and 7 depict the channel utilization of *link A* in case the standard IP model and in the case of our new model respectively. Furthermore, Table 2 summarizes the results indicating that the new model is better than the standard one in bandwidth consumption.

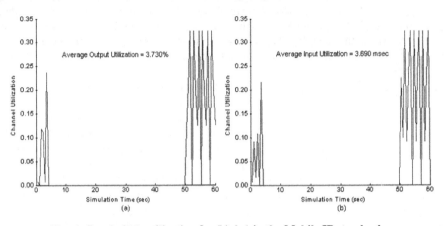

Fig. 6. Bandwidth utilization for *Link A* in the Mobile IP standard

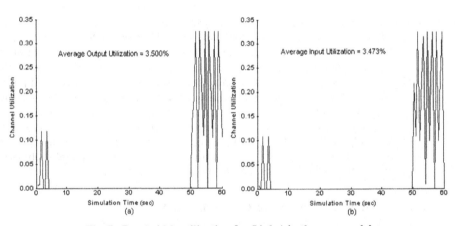

Fig. 7. Bandwidth utilization for *Link A* in the new model

Model	Link A		Total	Improvement
	In	**Out**		
Standard	3.690%	3.730%	7.420%	6.025%
New	3.473%	3.500%	6.973%	

Table 2. Improvement in bandwidth utilization for *link A*

As for the other home network connected via *Link B*, similar results are obtained and are represented in Figures 8 and 9 and summarized in Table 3.

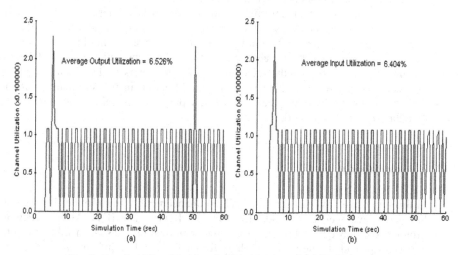

Fig. 8. Bandwidth utilization for *LinkB* in the Mobile IP standard

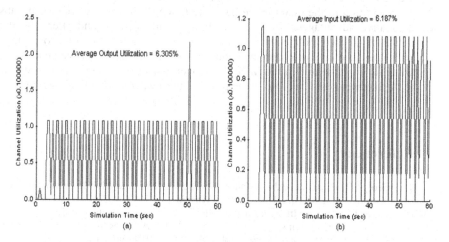

Fig. 9. Bandwidth utilization for *LinkB* in the new model

Model	LinkB		Total	Improvement
	In	**Out**		
Standard	6.404%	6.526%	12.930 %	3.387 %
New	6.187%	6.305%	12.492%	

Table 3.   Improvement in bandwidth utilization for *link B*

Anyhow, such numbers prove the fact that the new model surpasses the standard one. Indeed the improvement ratio could be large or small depending on the number of the mobile hosts and the way they communicate with other nodes. But, the results prove that there is a quantitative improvement. Moreover, it should be mentioned that all the parameters that have been applied throughout the simulation are taken as an assumption, with the fact that changing such parameters will definitely change the output numerical results. However, any modification in the simulation parameters does not contradict with the fact that the new proposed model is quantitatively better than the Mobile IP standard.

7 Conclusion

In this work, we have suggested possible enhancement to the Mobile IP protocol that has been developed and standardized by the IETF. The IETF Mobile IP working group has proposed a technique for route optimization. Based on this concept, this paper has provided a new methodology for caching mobile information. Also, a new vital component called the MIS has been added to the mobile architecture.

Simulation results supports the logical expectation of improved efficiency when the new architecture is used over the standard one.

In addition to the quantitative gain, the new model has achieved other substantial qualitative advantages. From a scalability point of view, and after describing the details of the two mobile architectures, it is quite clear that in order to deploy such a wide area caching mechanism; some sort of centralized management is required, which is implemented in the MIS. Moreover, our new design is transparent not only to the Internet hosts, but also to the protocol used whether it is IPv4 or IPv6.

References

1. George H. Forman and John Zahorjan.: The Challenges of Mobile Computing., Computer Science & Engineering, University of Washington, 1993.
2. Postel J.B., Editor.: Internet Protocol. IETF RFC 791, September 1981b.
3. P. Bhagwat, C. Perkins, and S. K. Tripathi.: Network Layer Mobility: an Architecture and Survey. *IEEE Personal Comm.*, Vol.3, No.3, June 1996, pp. 54-64.
4. Postel J.B., Editor.: Transmission Control Protocol. IETF RFC 793, September 1981c.
5. J. Ioannidis and G. Maguire Jr.:The Design and Implementation of a Mobile Internetworking Architecture. *In Proceedings of Winter USENIX*, San Diego, CA, January 1993, pp. 491-502.
6. John Ioannidis, Dan Duchamp, and Gerald Q. Maguire Jr. : IP-based Protocols for Mobile Internetworking. Department of Computer Science, Columbia University.
7. F. Teraoka, Kim Claffy, and M. Tokoro.: Design, Implementation and Evaluation of Virtual Internet Protocol. *In Proceedings of the 12th International Conference on Distributed Computing Systems*, June 1992, pp. 170-177.
8. F. Teraoka and M. Tokoro.: Host Migration Transparency in IP Networks. *Computer Communication Review*, January 1993, pp. 45-65.
9. Y. Rekhter and C. Perkins.: Loose Source Routing for Mobile Hosts. Internet draft, July 1992.

10. Charles Perkins, Andrew Myles, and David B. Johnson.: IMHP: A Mobile Host Protocol for the Internet. *Computer Networks and ISDN Systems 27*, December 1994, pp. 479-491.
11. Charles Perkins and Andrew Myles.: Mobile IP. *SBT/IEEE International Telecommunications Symposium*, Rio De Janeiro, August 1994.
12. Charles Perkins.: IPv4 Mobility Support. IETF RFC 2002, October 1996.
13. Charles Perkins. Mobile IP: Design Principles and Practices. Addison-Wesley, 1997.
14. Douglas E. Comer.: Internetworking with TCP/IP Volume I: Principles, Protocols and Architecture. Prentice-Hall, 1995.
15. Charles Perkins and David B. Johnson.: Route Optimization in Mobile IP. Internet draft, ftp://ftp.ietf.org/internet-drafts/draft-ietf-mobileip-optim-07.txt, November 1997.
16. S. Deering and R. Hinden.: Internet Protocol, Version 6 (IPv6) Specification. IETF RFC (1883), December 1995.
17. Charles Perkins and David B. Johnson.: Mobility Support in IPv6. Internet draft, ftp://ftp.ietf.org/internet-drafts/draft-ietf-mobileip-ipv6-07.txt, November 1998.
18. S. Thomson and T. Narten.: IPv6 Stateless Address Autoconfiguration. IETF RFC 1971, August 1996.
19. T. Narten, E. Nordmark, and W. Simpson.: Neighbor Discovery for IP Version 6 (IPv6). IETF RFC 1970, August 1996.
20. Scott O. Bradner and Allison Mankin.: IPng Internet Protocol Next Generation. Addison-Wesley Publishing Company, 1995.
21. Charles Perkins.: Mobile Networking Through Mobile IP. *IEEE Internet Computing*, February 1998, pp. 58-69
22. COMNET III User's Mannual, Planning for Network Managers Release 1.2, 1996.
23. Mockapetris P.: Domain Names-Concepts and Facilities. IETF RFC 1034, November 1987.

A Reliable Subcasting Protocol for Wireless Environments

Djamel H. Sadok, Carlos de M. Cordeiro and Judith Kelner

Centro de Informática, Universidade Federal de Pernambuco, Caixa Postal: 7851
Cidade Universitária, Recife, PE, Brasil
{jamel, cmc, jk}@di.ufpe.br

Abstract. This paper presents an end-to-end reliable multicast protocol for use in environments with wireless access. It divides a multicast tree into sub-trees where subcasting within these smaller regions is applied using a tree of retransmission servers (RSs). RM2 is receiver oriented [1] in that the transmitter does not need to know its receivers, hence offering better scalability. The Internet Group Management Protocol (IGMP) is used manage group membership whereas the IETF's Mobile IP offers support to user mobility and a care-of address (temporary IP address). Each RS has a retransmission subcast address shared by its members and which may be dynamically configured using IETF's MADCAP (Multicast Address Dynamic Client Allocation Protocol) [8]. Most importantly, RM2 uses a dynamic retransmission strategy to switch between multicast and unicast retransmission modes according to the extra load generated in the network as well as the wireless interfaces by packet retransmissions. It is shown through both analytical modeling and simulation that RM2's dynamic adaptation is not only important but necessary when considering mobile access.

1 Introduction: Multicast in Mobile Environments

The IETF defines two approaches to multicast at the IP level, namely *bi-directional tunneling* and *remote subscription*. Other techniques for unreliable multicast have also been adopted in [4, 5, 6, 7]. The solution adopted in [4], and later on refined in [5], has scalability problems and assumes that group membership is static, which is hardly true when considering mobile environments. [6] and [7] deals with problems such as tunnel convergence, but does not deal with packet loss and performance issues.

Commonly used reliable protocols include the scalable reliable multicast (SRM) [2] and the reliable multicast transport protocol (RMTP) [3]. On the one hand, SRM is based on an application level framework where it is the application's responsibility to guarantee packet sequencing. This protocol is compared to RM2 using simulation later in the paper. On the other hand, RMTP defines a hierarchy of designated routers (DRs), a concept also used in RM2. Although each DR is responsible for handling error recovery within a region of the multicast tree, it does not say whether this is done in multicast or unicast and, therefore, does not concern itself with the control of retransmission overhead. This is a serious drawback when considering the emerging mobile environments.

C.G. Omidyar (Ed.): MWCN 2000, LNCS 1818, pp. 174-185, 2000.

In this paper, a new reliable multicast protocol, called RM2 (Reliable Mobile Multicast), tailored for mobile environments is presented.

2 Specification of the RM2 Protocol

The role of RM2 is to take a stream of packets generated by an application and deliver it to all mobile as well as fixed hosts interested in receiving it in a reliable and optimized way. Furthermore, RM2 guaranties sequential packet delivery with no packet loss to all its multicast members.

RM2 assumes that the network is formed of multicast routers and that cells are big enough to allow users to join and leave multicast groups. The RSs (Retransmission Servers) perform selective retransmissions on the basis of feedback in the form of negative acknowledgements they receive. RM2 dynamically establishes the subcasting regions while taking into account retransmission costs.

2.1 The Role of a Retransmission Server

A multicast sender must first divide a multicast message into smaller fixed size packets (except the last one). To each one of these packets, RM2 associates a sequence number (n_{seq}). In order to guarantee end-to-end reliability, the receivers are required to send NACKs pointing out which packets they want to be retransmitted. In other words, RM2 implements selective packet retransmission. A NACK contains a sequence number N and bitmap B. N indicates that all packets with sequence number less than N have been correctly received by a given receiver. The bitmap B, on the other hand, shows which packets have actually been received. Consider the example where the ACK contains N=22 and B=01111101. In this case, the receiver wishes to indicate that it has correctly received all multicast packets with a sequence number less than 22 and that it is requesting the retransmission of packets 22 and 28 as indicated by the two 0s present in the bitmap B.

Sending NACKs to a multicast sender could lead to overwhelming it and causing an NACK implosion and even network congestion especially near the multicast source subnet. Therefore, RM2 divides the multicast network into hierarchical regions, where each one of these is controlled by a retransmission server (RS). This is responsible for gathering and processing NACKs from its region and for the retransmission of packets as requested by some receivers. Figs. 1(a) and 1(b) illustrate this concept. RM2 assigns RS functionality to fixed hosts selected on a network topology basis. The selection is subject to analysis in the following text.

The RSs are centric to RM2 support for reliable multicast sessions. Multicast packets are cached within the RSs buffers. The RSs are also responsible for the combination of NACKs received from lower RS hierarchies and hosts within their regions and responding to these when possible. If not, the retransmission request may be passed on to higher levels of RSs. Note that the separation between multicast routers and RSs is purposely done. It has the benefit of freeing router resources to the handling of multicast packets. RSs, on the hand, are better represented by hosts since they may need to use large buffering to keep copies of multicast packets for possible retransmission.

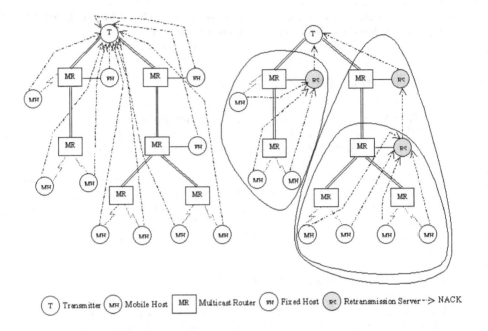

T) Transmitter (MH) Mobile Host [MR] Multicast Router (FH) Fixed Host (RS) Retransmission Server --> NACK

Fig. 1(a). Receivers sending NACKs to the multicast transmitter

Fig. 1(b). RS NACK processing

2.2 Establishing the RS Hierarchy

Core to the RM2 protocol is the establishment of a minimum cost spanning tree of RSs. The static selection of RSs in a WAN is on the basis of network topology. However, a multicast receiver selects dynamically its RS. Initially when there are no users in group a G, the RSs transmit a CFG_RS packet for G at each time interval T_{RS} in order to:

- Advertise its status as an RS to all potential members of a group;
- Define transmission costs where each CFG_RS packet contains a field with the cost of reaching the transmitter. This cost is null when the RS is not currently receiving data for G or that it is the actual transmitter. An example of a useful cost metric used in the simulation is the number of hops.

As soon as hosts join the multicast group they start receiving, in addition to multicast data packets, CFG_RS configuration packets. This way a host may be able to select the one that it is nearest to it on the basis on the cost information. The host then sends a REG_HOST to a suitable RS and sets a timer with a $2*T_{RS}$ value. The REG_HOST message has the following information:

- *HostType*: telling its RS whether it is a fixed or mobile host;
- *Group ID*: of the multicast group it is currently joining;

On receipt of a REG_HOST message a RS invokes the following actions:

- The RS updates its fixed/mobile host counters within its region and registers the hosts' unicast address;
- In the event that this is the first host to enter a multicast group at this RS, it dynamically allocates a D (multicast type) address, for example G1, using a protocol such as MADCAP [8]. This is the retransmission address for all receivers at this RS. Next the RS sends a RETR_ADDR packet telling its new host which multicast retransmission address its has to listen to. From now on, a new group G1 with a subcasting address has been established at this RS;
- Based on the number of retransmission requests a RS receives within a region, it may retransmit using either unicast or subcasting. This flexibility is a clear advantage over other protocols such as RMTP and SRM. Details of this mechanism are presented later in the paper. With the insertion of new hosts, new regions with separate RSs are formed. Each new RS performs the same procedure as a host in order to get a retransmission feed from a higher level RS. RM2 uses the Djikistra algorithm to establish the spanning RS tree where hop count is cost metric.

2.3 Support for Fixed Receivers

At the network level IGMP is used for group management. Existing multicast routing protocols such as the distance vector multicast routing protocol (DVMRP), protocol independent multicast (PIM) and multicast OSPF (MOSPF) may be used in conjunction with RM2 for routing multicast packets. A RS distinguishes between mobile and fixed hosts using the *HostType* field present in the REG_HOST registration message. Similarly, a host may leave a retransmission subcast group at a RS by sending it a LEAVE_GROUP message. The RS needs to keep track of the number of hosts in its region and should remove its link to the upper RS when it has no members to serve.

2.4 Support for Mobile Receivers

Mobile devices are somehow more delicate to handle at the RS. MHs send a normal IGMP join message to a multicast router (MR) using the care-of address from a foreign network (FN) obtained through the use of the IETF's mobile IP. The MR uses normal procedure to include itself in the IP multicast tree. The MH then sends a REG_HOST to its closest RS with the *HostType* field set to mobile. The following situations may happen as a result of host mobility:

- A MH sends an IGMP leave group message to its current MR before performing handoff. The MR may then check through a new IGMP query message to see if there are still multicast host members in its subtree. If not, it would remove itself from the multicast tree. Furthermore, the MH also sends a RM2 LEAVE_GROUP message to its respective RS in order to be removed from the retransmission subcast list;

- A second situation rises when the MH is unable to leave the group before handoff is complete or that the "leave" message is lost. Therefore, the MR may only know if there are hosts left at its subtree only after it sends a new IGMP membership query message. Since the IGMP leave message may itself be lost be lost, a registration timeout is associated to each MH. The RS relies on NACKs sent by MHs to know which of these remain active in its region. An interesting scenario emerges when a MH performs a handoff and moves to a new network (at another FA). This leads to packet loss due to the fact that the MH is unable to receive multicast packets when it is in between FAs. At reconnection, the MH must send an RM2 REQUEST message to start receiving from where it left the multicast transmission before the handoff.

2.5   Support for Mobile Transmitters

Reconfiguring the entire multicast tree each time a sending MH moves to a new FN is costly. Another problem results from the sending of NACKs to a FN where the MH no longer is. RM2 adopts two approaches to deal with these problems:

- When the MH is within its home network, it performs a link level multicast. The HA then forwards these downstream;
- If the MH is visiting a FN, it tunnels packets to its HA which then forwards these using local links and WAN interfaces. In other words, the HA in this case performs the role of a gateway. For performance and scalability reasons, RM2 limits the use of IP tunnels when there is a multicast mobile source transmitting.

2.6   RM2 Error Recovery

NACKs are used by hosts to signal lost and corrupt packets. RM2 adopts a similar approach to RMTP in that it collects NACKs per packet within a queue during a specific time interval and then retransmits the requested packets. This scheme clearly needs further tuning in the case of mobile hosts, especially considering that the requesting mobile hosts may have moved on to new FNs since they sent NACKs when dealing with high mobility users for example. RM2 adopts a dynamic retransmission technique at the RS level as shown in the following two scenarios:

- High error rate: representing environment with a high number of mobile users with wireless access;
- Low error rates: representing situations with a relatively low mobile to fixed users ratio;

A RS continuously monitors the number of fixed and mobile users and their NACKs. It is shown when describing the analytical model for RM2 that there are two main mathematical restrictions that control the adaptation of the retransmission algorithm that decides whether retransmitted packets at a RS are sent to all group members via multicast (more specifically subcast) or that they are sent via unicast to all those hosts that requested them. When the number of NACKs is relatively low, unicast is used at the RS to retransmit packets. RM2 monitors the network load, as shown in the analytical model, and should this becomes relatively high, it would then switch to multicasting retransmission packets. However, it is important that RM2 does not attempt to reduce network load at the detriment of low speed wireless interfaces. Indeed, RM2 only uses multicast as long as the wireless channel occupation remains bellow a threshold. Between overloading the fixed network and the wireless links, RM2 chooses to limit the traffic on the latter. There are, however, situations where, even when unicast is used at the RS, the wireless interface may see too many retransmission packets. The RS may then send, on behalf of its MHs, a REDUCE_FLOW message to the multicast source. Note that it is not mandatory for a transmitter to comply with this message. Table 1 shows a summary of RM2 messages.

Table 1. RM2 messages

Message	Originator	Receiver	Description
DATA	Transmitter	Receivers & RSs	Data Packet
REQUEST	Receivers & RSs	RSs	Requests the retransmission of lost packets
REPAIR	RSs	Receivers & RSs	Retransmission of lost packets
CFG_RS	Transmitter & RSs	Receivers & RSs	Building RS tree and establishing subcast regions
REG_HOST	Receivers and RSs	RSs	Host registration
RETR_ADDR	RSs	Receivers & RSs	Informs a region's retransmission subcast address
LEAVE_GROUP	Receivers & RSs	RSs	Indicates that a host is leaving a group or simply doing a handoff.
REDUCE_FLOW	RSs	RSs	Flow Control

3 Modeling Packet Retransmission

A n-ary network topology with a depth h is considered with equal fixed link costs and where retransmission requests are assumed to be uniformly distributed. If PS denotes the packet size, then equations (1) and (2) give the unicast and multicast retransmission costs to K receivers respectively.

$$C_U(K, h, PS) = K \times h \times PS . \tag{1}$$

$$C_M(n, h, PS) = n \frac{n^h - 1}{n - 1} PS , n > 1 . \tag{2}$$

The tree depth h depends is related to the total number of hosts (N_T) as shown in (3):

$$h(N_T, n) = \log_n \left(\frac{N_T(n-1)}{n} + 1 \right) . \tag{3}$$

Define $P(Xf_{i,j})$ and $P(Xm_{i,j})$ as the probability that a fixed/mobile host i requests a retransmission of packet j. They are respectively given by:

$$P(Xf_{i,j}) = 1 - (1 - P(E_F))^h$$
$$P(Xm_{i,j}) = 1 - (1 - P(E_F))^{h-1} (1 - P(E_M)) .$$

where E_F and E_M are packet error rates for fixed and wireless links.

Let $P(Xf_{i,j})$ and $P(Xm_{i,j})$ be the probabilities that a fixed or mobile host i requests the retransmission of packet j respectively. Furthermore, let $P_i(Y = K_F, Z = K_M)$ represent the probability of having exactly K_F and K_M retransmission requests of a given packet i for both fixed and mobile hosts, while N_F is the total number of fixed hosts and N_M the total number of mobile hosts. Since retransmission request events are independent and that $P(Xf_{i,j})$ and $P(Xm_{i,j})$ remain practically constant during the experiment, $P_i(Y, Z)$ may be modeled as a binomial distribution. Therefore, we have:

$$P_i(Y = K_f) = \binom{N_F}{K_f} P(Xf_{i,j})^{K_f} (1 - P(Xf_{i,j}))^{N_F - K_f} . \tag{4}$$

for fixed hosts, and

$$P_i(Z = K_m) = \binom{N_M}{K_m} P(Xm_{i,j})^{K_m} (1 - P(Xm_{i,j}))^{N_M - K_m} . \tag{5}$$

for mobile hosts.

Through consecutive mathematical refinements of equations (4) and (5), we obtain equations (6) and (7) as representing packet retransmission in both unicast and multicast modes:

$$E(CR_U) = hPS(N_F P(Xf_{i,j})(1 - P(Xf_{i,j})) + (N_F P(Xf_{i,j}))^2 + \tag{6}$$
$$N_M P(Xm_{i,j})(1 - P(Xm_{i,j})) + (N_M P(Xm_{i,j}))^2) .$$

$$E(CR_M) = n \frac{n^h - 1}{n - 1} PS(N_F P(Xf_{i,j}) + N_M P(Xm_{i,j})) . \tag{7}$$

Fig. 2 presents the network load behavior as the number of retransmission requests arriving at the transmitter increases (equations (i) and (ii)). We see that, as the number of requests grows, the overload generated by the retransmitted packets depends on the retransmission mode (unicast or multicast). Ideally, RM2 should change into multi-

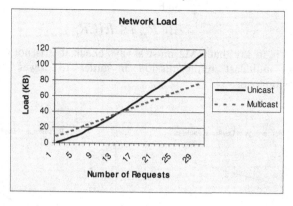

Fig. 2. Network Load

cast retransmission when the two curves intersect as shown in fig. 2. Table 2 shows some of parameters used for this analysis. Through the variation of the ratio mobile/fixed hosts, we show that the effect of mobile hosts is far more important on retransmission load and that the initial unicast retransmission scheme cannot be maintained when there is a large network.

Table 2. Some Parameters for the Analytical Evaluation

Parameter	Value
$P(E_F)$	10^{-9}
$P(E_M)$	10^{-3}
n-arity	2
PS (Packet Size)	1KB

Actually, the RM2 retransmission mechanism takes into account two restrictions: R1 and R2. R1 guaranties that the network load does not exceed an established threshold p whereas R2 ensures that wireless retransmission channel utilization is not overwhelmed by multicast retransmitted packets. R1 is given below (equation 8):

$$\alpha.E(CR_U) \leq p . \tag{8}$$

Where α:

$$\alpha = \frac{1}{N_M} \cdot \frac{N_M P(E_M)}{N_M P(E_M) + N_F P(E_F)} = \frac{P(E_M)}{N_M P(E_M) + N_F P(E_F)} . \tag{9}$$

In other words, α represents the average error rate related to fixed and wireless errors as well as the number of both user types. Fig. 3 shows the impact of both fixed (N_F) and mobile users (N_M) (equations (vi) and (vii)). Through fig. 3, we see that mobile users have a much bigger effect on the load generated by duplicate packets than fixed users. Therefore, RM2's retransmission mechanism was specially adapted to deal with mobile users.

R2, on the other hand, is given by:

$$E(CR_M) \leq E(CR_U) .\tag{10}$$

That is to say that RM2 must always check if it is not time to switch from a unicast to multicast retransmission in order to lower the network packet retransmission load.

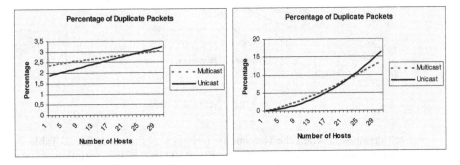

Fig. 3. Impact of Fixed and Mobile hosts

4   RM2 Simulation

RM2 was implemented in the U.C. Berkeley/LBNL ns-2 simulator. The source code is available from *http://www.di.ufpe.br/~cmc/research/rm2/rm2-source.zip*.

Topology. A backbone topology with 7 multicast routers has been used with 2 Mbps links and 10ms delay. The scenario consists of a maximum of 50 users joining a multicast transmission at different levels of the hierarchy. Fixed access is characterized by a 10Mbps speed and 5ms delay, whereas mobile access uses 14 kbps and a 50ms delay. We used ns-2 DVMRP (Distance Vector Multicast Routing Protocol) for building routing tables, IGMP for group management and mobile IP for address allocation.

Transmission. Each simulation experiment starts at time 1 second, and after the bootstrap phase of 8 seconds a CBR (constant bit rate) source starts transmitting 5000 1KB packets at 64 Kbps.

Error Model. In RM2's simulation, errors are a result of buffer overflow in routers and transmission packet error rates. Whereas the first one represents the dominant source for Internet packet loss, the second one reflects the error probability inherent to each link, as explained in the analytical model and configured in table 2. As presented earlier, the analytical model determines that the loss rates receivers experience are obtained by compounding the loss rates on the links from the sender to the receivers.

Results. The results mainly validate the analytical modeling earlier shown. The network load shown in fig. 4 presents similar tendencies to the one resulting from the analytical model presented fig. 2.

Similarly, the results show that the impact (see fig. 5) of both fixed and mobile users is in line with the one from the analytical model shown in fig. 3.

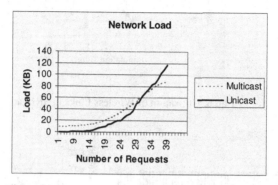

Fig. 4. Simulation of Network Load

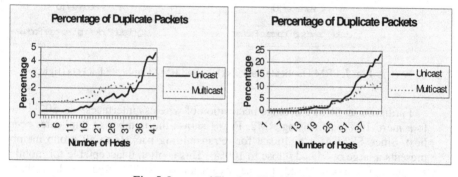

Fig. 5. Impact of Fixed and Mobile Hosts

5 Comparing RM2 Performance with SRM

Finally, we compare RM2 and SRM performance under similar loads. SRM has been selected since it is well referenced in the literature and readily available in the ns-2 simulator. We set R1 as a condition for RM2 to ensure that wireless interface may not see more than 20% of retransmission. Fig. 6 illustrates link utilization during the simulation.

The results are somehow intuitive, SRM generates higher levels of network occupancy since it always retransmits in multicast mode which may of course overload low speed wireless interfaces (in fig. 6, there are times where SRM uses almost 80% of this capacity).

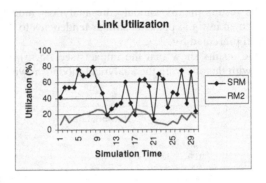

Fig. 6. Utilization of Wireless Links in SRM and RM2

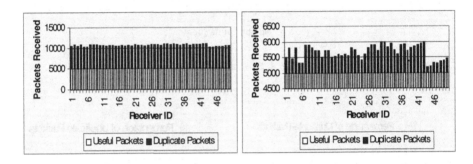

Fig. 7. SRM Duplicate Packets (left) and RM2 Duplicate Packets (right)

Furthermore, SRM's average occupation of wireless interfaces was 50% compared to a mere 16% when using RM2. Fig. 7 shows the number of duplicates seen by a host. Since SRM uses multicast for retransmitting packets to all group members, it presents a large overhead (close to 100%). This is often unacceptable for mobile hosts specially when compared to the overhead of RM2 which is less than 20% (maintaining restriction R1). Clearly, RM2 retransmission adaptation not only optimizes network traffic load but more importantly spares low speed wireless links.

6 Conclusions

This work has presented a new reliable multicast protocol, suitable for use in wireless environments. RM2 defines a hierarchy of retransmission servers which implement subcasting. It was shown through analytical modeling and simulation that the adaptation mechanism reduces multicast traffic due to packet retransmission and saves link utilization at the wireless interface. Finally, through simulation, RM2 efficiency over SRM was shown to be superior.

References

1. S. Pingali, D. Towsley, J. Kurose, "A Comparison of Sender-Initiated and Receiver-Initiated Reliable Multicast Protocols", *Proc. ACM SIGMETRICS Conf. On Measurement and Modeling of Computer Systems*, May 1994
2. S. Floyd, V. Jacobson, S. McCanne, C. Liu, L. Zhang, "A Reliable Multicast Framework for Light-Weight Sessions and Application Level Framing", *ACM SIGCOMM'95, Conf. on Applications, Technologies, Architectures and Protocols for Computer Communications*, August 1995
3. S. Paul, K. Sabnani, J. Lin, S. Bhattacharrya, "Reliable Multicast Transport Protocol (RMTP)", IEEE Journal on Selected Areas in Communications, April 1997
4. A. Acharya, B. Badrinath, "Delivering Multicast Messages in Networks with Mobile Hosts", *Proc. Of the 13th International Conference on Distributed Computer Systems*, May 1993
5. A. Acharya, B. Badrinath, "A Framework for the Delivery of Multicast Messages in Networks with Mobile Hosts", *Wireless Networks*, 1996
6. V. Chikarmane, C. Williamson, R. Bunt. W.Mackrell, "Multicast Support for Mobile Hosts Using Mobile IP: Design Issues and Proposed Architecture", *ACM/Baltzer Mobile Networking and Applications*, 1997
7. T. Harrison, C. Williamson, R. Bunt. W.Mackrell, "Mobile Multicast (MoM) Protocol: Multicast Support for Mobile Hosts", *Department of Computer Science*, University of Saskatchewan, Canada
8. MADCAP Protocol, http://www.ietf.org/internet-drafts/draft-ietf-malloc-madcap-07.txt

Author Index